The Renewable Energy Alternative

THE
RENEWABLE ENERGY
ALTERNATIVE

How the United States and the World Can Prosper

Without Nuclear Energy or Coal

JOHN O. BLACKBURN

Duke University Press   Durham 1987

Contents

List of Tables and Figures

Preface

This book is for people who still think that there are energy problems ahead, and who are not satisfied with an energy future dominated by coal and nuclear power.

Fortunately, the time has come when one can examine other alternatives seriously—in particular, energy conservation and renewable energy—without being relegated at once to the outer fringes. Southern California Edison, Pacific Gas and Electric, and numerous other progressive utilities, having embraced these alternatives, are a bit too formidable to be drummed out of the corps. The Office of Technology Assessment, the Gas Research Institute, and the Harvard Business School likewise are not thought of as outposts of a lingering counterculture.

Still, not enough people in the energy supply industries, in government, or in private agencies concerned with energy policy are aware of the full range of possibilities in energy conservation and in the numerous renewable energy technologies. For them, this book is an attempt to lay out, in an ordered fashion, the basis in fact, logic, and judgment, for an energy-efficient, renewable energy future.

Many other people could have written this book, and several of them nearly did. A great deal is now known about the large potential for energy conservation and for gains in energy efficiency. Many people are involved in that work, and know more about it than I do. Likewise, a great deal is now known about the large renewable energy potential. Many people are working on these technologies and their knowledge is expanding by the day.

Fewer people have brought these strands of energy analysis together. In the United States, for an overall perspective on energy supplies and demands, one thinks of Denis Hayes, Amory Lovins, or the Union of Concerned Scientists.

Lovins was among the first to sketch out the transition to a renewable energy future. He has not published a quantitative study for the United States—perhaps because it seemed so painfully obvious to one informed so early and so well.

Hayes's work in the 1970s did not have the benefit of more recent discoveries of the truly large size of potential conservation-efficiency gains. During his term as director of the Solar Energy Research Institute (SERI), he oversaw the preparation of the most thoroughgoing study of the U.S. conservation-efficiency potential yet done. That study also made estimates of potential renewable energy outputs in the United States for the year 2000.

That study made no claims about any renewable energy future; it merely documented potential savings achievable by the year 2000. That was too bold for its time; it did not get through the review process for publication during the last year of the Carter administration, and it had no appeal at all to the incoming Reagan administration. U.S. Representative Richard L. Ottinger from New York may be credited with bringing the study to the public.

From the perspective of this book, the SERI study shows how to get halfway (geometrically) to a renewable energy future in twenty-three years (1977–2000). Hayes could easily have written about the other half. Lovins and Hayes are but two examples. There are others.

When I moved to Florida in 1980, it seemed that the case for a long-run energy-efficient, renewable energy future had already been made, then most recently by the Union of Concerned Scientists. The remaining task seemed to be one of working out the details for non-typical areas like Florida.

I took up that task, the results of which will soon be ready for publication. While that project was under way, it became increasingly clear that very few people in government, in business, the public, or around what are loosely called "energy policy circles," yet saw what Lovins, Hayes, and others in the vanguard saw so clearly.

And so, to paraphrase President Reagan, here we go again. Once more, with numbers, with the benefit of recent experience, and with a few original ideas, we lay out the case for an energy-efficient, renewable energy future as a viable alternative.

Among the friends and colleagues who read various drafts and offered useful comments I am especially grateful to Allen Kelley, Arjo Klamer, Jan Beyea, Peter Hess, and Craufurd Goodwin. They do not necessarily share my conclusions, and they should not be implicated in any remaining errors. I am grateful to Chip Weston for the graphics and to Marilyn Ross who typed the final manuscript with competence, speed, and considerable damage to a Christmas holiday.

I

The Energy Debate:

More Divergence than Consensus

The oil market upheavals of 1973 brought energy matters onto the public agenda for a decade. Public concern has faded as surpluses have appeared in markets for conventional energy supplies, and as oil and gas prices have dramatically declined. One would suppose that any problems that may have existed have been solved.

The most interesting feature of the American energy debate is that no consensus has been reached in the years since 1973. Initially, it seemed clear what had to be done. Then, sharply different views emerged; first, as to the nature of energy problems, and then, as to appropriate policies to address those problems. Those who still concern themselves with energy matters are strongly divided to this day.

One cluster of views about energy may be characterized as "supply-oriented." When the 1973 oil crisis erupted, the nation turned naturally to its energy experts. At that time most of the people who were informed about energy matters were in the energy supply industries or in government agencies with energy-related responsibilities—the Atomic Energy Commission or the Department of the Interior.

The response to the oil shortage was quite expectedly framed in terms of increasing energy supply, especially in forms other than oil. The basic premises motivating supply-side arguments included the following assumptions:

—Energy demand will grow with output. A healthy and growing economy requires a growing supply of energy. In early formulations the rate of energy supply growth was about equal to the growth rate of the economy.

—With proper incentives, some increases in domestic oil and gas outputs would be possible, at least for a while.

—Reliance on coal and nuclear power would have to expand rapidly. For longer periods during which conventional oil and gas outputs would level off or decline, still more coal and nuclear power, along with synthetic liquid fuels, would be needed to accommodate growth and to replace oil and gas.

—Electricity use would continue its long-established pattern of growing more rapidly than total energy use. This assumption fit neatly with the expected rapid growth of nuclear power.[1]

—Solar energy (or, as we now say, renewable energy)[2] might someday provide significant amounts of energy, but not within the twentieth century.

—These very large increases in energy supply might pose difficult economic and technical challenges, but such challenges would have to be overcome. Similar challenges had been met before. Lead times for the energy supply investments were long, so the process had to begin soon to avoid shortages in the near future.

A view something like that just described was held by virtually every national government, by the energy supply industries, and by international institutions such as the World Bank and the International Energy Agency.[3]

There are other views, to be sure; they will be examined in a moment. The view just enunciated, only somewhat modified by the experiences of more than a dozen years, is still the most widely held in official circles. One might characterize it as the "establishment" view.

Economists have tended to take a somewhat different approach to energy analysis. If petroleum were to become scarce, its price would rise. It would be used more sparingly and other fuels would be substituted for it. If energy prices in general were to rise, other inputs (capital, for example) would be substituted, and technological change would be directed toward more energy-serving processes.

These rather abstract conclusions, which follow from widely accepted premises in the discipline, were reinforced by some important empirical work in the economics of exhaustible resources.[4] New sources of raw materials were continually found, advancing technology brought lower-grade and less accessible sources into markets, and substitutes appeared. Real prices showed no upward trend over long periods of time.

The general lack of concern among economists about exhaustible resources (metal ores and other minerals) rested in part on the implicit assumption that increasing quantities of energy would be available. When pressed by those more concerned about resource exhaustion, economists usually suggested that nuclear power was the backup energy source.[5]

Many economists see no need for an "energy policy"; market forces can determine the rate at which oil and gas are used up and substitutes brought into use. At most, goverments might be concerned with the external effects of various energy sources and, perhaps, with basic research. For externalities, such as air pollution from burning coal, marketlike remedies to internalize costs would be preferred to highly prescriptive regulations.

The most market-oriented economists would be relatively indifferent to the amount of energy to be used in 1990 or 2000 or at any time, or whether it was supplied by oil, coal, nuclear power, or solar sources. Market forces would settle these issues.[6]

Other economists, even as they continue to emphasize the importance of price-led adjustments, see a larger role for policy measures.[7] They cite misleading price signals, the lack of knowledge appropriate to changed circumstances among millions of energy users, the selection of energy-using devices by those who do not later pay for energy consumption (builders, landlords), and time horizons for investment decisions that differ from those appropriate to the larger society.

As the decade of the "energy crisis" progressed, other views came forward. Solar energy advocates began to be heard, and solar energy research support began to appear in federal energy budgets. Estimates of large potential output from various forms of solar or renewable energy began to appear. Proponents of all-solar futures were not numerous, for they faced several handicaps. One was the size of energy "requirements" or "demands" to be met. It was still commonly thought that energy outputs would need to double from 1973 to 2000; drawing up all-solar scenarios for such high levels of energy use was a formidable task.

The most important developments were taking place in still another arena. Investigations into the manner in which energy was actually used in its "end uses"—in buildings, in transportation, and in industry—were beginning to produce large estimates of potential energy conservation. One started with the "end uses" of energy and then aggregated to total energy demands, rather than working only with aggregates. If one looks at space heat, hot water, lights, cooling, auto transport, copper smelting, and all the other ways in which energy becomes useful to its buyers, a host of new possibilities appears. One can ask for each use whether there are cheaper alternatives than simply using more and more energy as the economy grows.

These investigations suggested so many ways of reducing energy use without inconveniencing anyone—ways that were actually more economical than new supply at higher energy prices—that some writers began to speak of the "conservation revolution."[8] In area after area it became apparent that increasing the efficiency with which energy was used would be less costly than supplying new energy.

One of the earliest studies that pointed to a large conservation potential was done by the Ford Foundation's Energy Policy Project.[9] Its "Zero Energy Growth" scenario projected that energy growth could level off around 1985. This study was roundly attacked by the energy supply industries and by some economists for its lack of "rigor."[10] The authors themselves paid more attention to a higher-energy "technical fix" scenario since they assumed that changes in living patterns would be required to bring about "zero energy growth." As things have turned out, U.S. energy use is lower than in the "zero growth" scenario, but without life-style changes.

This emerging body of research was sufficiently far along to find an outlet in a National Academy of Science study. The Committee on Nuclear and Alternative Energy Systems published its report in 1979.[11] Its scenarios, with varying combinations of economic growth rates and energy prices, produced a very wide range of results. Energy use in the year 2010 could be less than in 1975, for instance—or more than twice as much. Soon after the report appeared, the price of oil jumped again—over halfway to the study's price in the year 2010—a price that would result in a very low growth rate in energy use.

In rapid succession other studies appeared that emphasized the large conservation-efficiency potential and the relatively low cost of the measures to exploit that potential. These studies are cited in chapter 2. It suffices here to note that an important new element had been injected into the American energy debate.

This line of research was reinforced by the data on energy use in the United States and elsewhere in relation to Gross National Product (GNP).

Energy intensities of economies were dropping in most nations—by 10 to 20 percent in the 1970s alone.

Still the "establishment" or supply-oriented view of energy futures continues to dominate thinking in governments, in most of the energy supply industries, and in most international institutions. Expected energy growth rates are somewhat lower than in the 1970s, and they start from lower bases; this much reflects recent experience. In all other respects the features outlined here continue to characterize "official" energy policy.[12]

The persistence of this "official" view may be explained in part by the long-term career commitments of people in government and industry. It is not easy to make radical changes in these matters when one's conceptual framework must be discarded. In other cases, large financial commitments have already been made. Only the sturdiest can admit to and write off a $1 billion mistake even if doing so would avoid a $5 billion write-off later, on someone else's watch.

Then there is the "real men don't eat quiche" phenomenon. "Real" physicists may work on breeder reactors or fusion power, but not on the physics of heat transfer in buildings or on stove tops. Fortunately, some physicists have overcome this barrier.

The view presented in this study is a combination of the conservation and renewable energy views sketched earlier. It brings them more fully into a market framework. It takes advantage of all the material now available on the size of the conservation potential. It also takes advantage of a decade's experience with renewable energy technology.

There are many studies of the large potential for conserving energy or using it more efficiently; there are other treatments of renewable energy. Only in a few cases, however, have these strands been brought together into a renewable energy, energy-efficient view of the future. One of the first investigators to make this connection explicit was Amory Lovins. His influential 1976 article in *Foreign Affairs*[13] helped add new dimensions to the energy debate. He chose not to make a quantitative demonstration for the United States, but he did sketch a transition path. Since Lovins's study was done in the mid-1970s, further energy demand growth was assumed, as was an interim increase in the use of coal.[14] A similar sketch was made by Denis Hayes;[15] at that time it appeared that the maximum conservation potential was 40 to 50 percent of energy use. One quantitative study was published by the Union of Concerned Scientists in 1980.[16] The authors projected energy demands to the years 2000 and 2050 using an estimated annual percentage gain in energy efficiency. They then showed that the projected demand could be met from solar sources. Bent Sorensen also published a renewable energy analysis for the United States in 1980.[17] A low-energy study with life-style changes and some coal use was done by John S. Steinhart and colleagues.[18]

The present study may be viewed as a continuation of these efforts. It benefits from recent developments in energy conservation technology and renewable energy, as well as from the practical experiences of the intervening years. The long-run energy demand figures shown here are well below those of the Union of Concerned Scientists' analysis, since they are based on more recent estimates of the conservation potential. That aspect of the analysis makes it easier to meet energy demands from renewable sources and diminishes the size of the energy storage problem that arises from reliance on intermittent sources.

At this point I also should mention a study with considerable impact: *Energy Future*,[19] by a faculty group at the Harvard Business School. It still spoke of conservation gains in the 40 percent range, not the larger ones given here. It was decidedly positive on the role of solar or renewable energy.

The most complete and up-to-date work on renewable energy supply is *Renewable Energy* by Daniel Deudney and Christopher Flavin of the Worldwatch Institute.[20] Their work concentrates on the large potential output of renewable energy in the world and therefore does not dwell heavily on the conservation–energy efficiency aspects of energy analysis.

In spite of these and other efforts, and in spite of the experience of the last dozen and more years (during which there has been little growth in world energy use), no senior government officials in any nation, or their counterparts in industry and the international institutions, think of energy futures largely in terms of energy efficiency and renewable energy. The "establishment" view is alive and well.

The public and these key decisionmakers very much need to know that we do indeed have choices—choices that are more attractive economically than high-energy, high-coal, and high-nuclear futures, and more attractive in all other respects as well. Hence this additional effort, this carrying forward of the demonstration.

This study proceeds by asking and then by answering in turn the following questions:

1.What would U.S. energy use in 1980 have been if all cost-effective investments in energy efficiency had already been made?[21] (Cost-effectiveness is defined with respect to long-run marginal supply prices of new conventional energy supplies.) This question is explored in some detail in chapter 2. The answer—that energy use would have been lower by two-thirds or three-fourths—is documented in detail.

2. Is there enough renewable energy potentially available in the United States to meet that reduced energy demand at prices no greater than those cited above? Chapter 3 reviews the actual and potential supplies of renewable energy in the United States. It deals only with sources known to work technically—with actual operating experience. The criterion for economic feasibility is that each renewable energy source considered must already be competitive with conventional energy sources or must have good prospects of becoming so within a decade as indicated by reputable engineering-economic studies. With this criterion the requisite amounts of renewable energy are forthcoming in the United States.

3. Since output grows over time, energy demands also may grow. Can growth be accommodated in the above scheme? This matter also is discussed in chapter 3.

4. A demonstration for the nation as a whole may leave major questions unanswered for states, cities, and regions. Chapter 4 addresses these issues. It shows that renewable energy sources are much more evenly distributed than conventional ones. All areas can be at least partially self-sufficient; trade in electricity and liquid fuels would take place at much lower levels than is now the case, and that trade would balance out regional supply-demand discrepancies.

5. The presentation of ideas has proceeded to this point as if nuclear power and coal did not exist. Chapter 5 investigates the prospects for coal, synthetic fuels, and nuclear power.

6. It is one thing to show that the United States, a well-endowed and not densely populated nation, has a credible renewable energy future. That tells us nothing about other industrial nations and Third World states. What are their prospects? Chapters 6 and 7 go into these matters—with positive findings.

7. If all this is true, then, can we get there from here? What are the implications for energy policy? How does one proceed when falling oil and gas prices work strongly against further efficiency gains or new energy supplies of any kind? These matters are treated in chapters 8 and

9. Policies matter, and there are many things worth doing without regard to this year's price of oil. The main object of energy policy, as it is developed here, is to create a fair set of rules so that conventional energy supply, conservation investments, and renewable energy supplies can compete on equal terms. In every institutional setting in which this state of affairs has been approximated, the message of this book has been vindicated: energy efficiency and renewable energy emerge as the technologies of choice.

2

The U.S. Energy

Conservation-Efficiency Potential

This chapter explores the potential for using energy more efficiently in the United States and finds that potential to be large. Nineteen eighty is used as a reference year;[1] energy use is examined in different sectors (buildings, transportation, industry) and for different end uses (such as space heat, lights, process heat, and freight transportation).

This is not a discussion of deprivation or reduced living standards; rather, it is an exploration of means to deliver the same "energy services" (warm rooms, cold refrigerators, needed tons of steel produced) with less energy expended. The conclusion is clear: energy use, for the same services provided in 1980 in the United States, could have been smaller by two-thirds or three-quarters. This conclusion is supported by considering, in turn, each of the significant end uses of energy and the economical means now available to reduce energy usage.

If energy use is to be reduced by such significant amounts, one has to ask, "at what cost?" The answer depends on the cost of new energy supplies. Actions or investments to increase the efficiency of energy use are warranted if they save energy at unit costs less than new energy supplies. This criterion for measuring the energy-saving potential of the economy requires some estimates of the costs of new energy supplies, or, in terminology familiar to economists, the long-run marginal supply prices of oil, gas, coal, and electricity.

After the conservation-efficiency estimates are made, they are compared with those found in other recent studies. The estimates given here are larger than most, but they are not necessarily inconsistent. Briefly, this study uses estimates of long-run marginal supply prices that are higher than prices actually seen by energy buyers in recent times, whereas other studies typically use current prices. This study, in contrast to other studies, also allows a longer period for adjustment.

Most economists think about quantity changes in response to price changes in terms of elasticity coefficients. The estimates of energy-use reductions are therefore compared with results obtained by applying price and income elasticities from the literature. Again, the estimates here are larger, and the reasons for this difference are explored in detail.

Finally, U.S. energy-use data from 1973 to 1985 are examined. Energy use relative to GNP fell about 25 percent over that period. This is partly a result of the failure, by 1985, to return to a "high-employment" level of output and of the relative decline of the energy-intensive heavy industry sector. Nevertheless, when adjustments are made for these factors, the decline in energy use relative to output is still large—a result quite consistent with the existence of a large conservation-efficiency potential.

U.S. Energy Use in 1980

The first step we must take, then, is to examine energy use in the United States in 1980. That use was about 76 quads of primary energy. A quad (quadrillion British thermal units) is equivalent to the energy content of about 500,000 barrels of oil per day for a year, or about 40 million tons of coal, or the output of sixteen large power plants running normally all year. "Primary energy" is measured before energy losses in conversion are deducted. Such conversion losses take place, for example, when crude oil is refined into gasoline or when coal is burned to produce electricity.

Custom now divides the economy into three large energy-using sectors: residential and commercial, transportation, and industrial. The latter sector is broadly defined to include agriculture, mining, and construction. Table 1 shows how much energy was used in each sector in 1980 and, in percentages, the purposes for which it was used.

A quick look at table 1 highlights the importance of heat at various temperatures in the nation's energy budget. Space heat, hot water, industrial process heat (and, in categories not shown separately, cooking, crop drying, and clothes drying) take some 40 percent of the nation's

Table 1 U.S. Energy Use, 1980, by Sector and by End Use Within Sectors.

Sector	Energy use (quadrillion BTU's)	End uses (Percentage within each sector)	
Residential and commercial structures	25.7	Space heat	46
		Space cooling	13
		Lights	12
		Hot water	9
		Refrigerators	6
		Other	14
			100
Industrial (includes agriculture, mining, and contruction)	30.5	Process heat	43
		Temperature: High	(17)
		Medium	(21)
		Low	(05)
		Mechanical drive	26
		Electrolysis	5
		Metallurgical coal	7
		Feedstocks and asphalt	11
		Other	8
			100
Transportation	19.7	Autos	45
		Light trucks	13
		Freight trucks	13
		Aircraft	12
		Rail	3
		Water	10
		Pipelines	3
		Bus	1
	76.0		100

Sources: Monthly Energy Review, Solar Energy Research Institute, *A New Prosperity,* Audubon Energy Plan, 1984.

energy. Transportation, which means with very few exceptions liquid fuels, takes another 25 percent. The large role of heat and liquid fuels in overall energy demand will be recurring issues. In an effort at simplicity, table 1 omits some important information. When electricity is generated, about two-thirds of the primary energy involved is lost at the power plant as waste heat. This loss, a conversion loss, amounted to about 16 quads in 1980 but is allocated to the three sectors according to their use of electricity. If one adjusts for this and for other conversion losses, and looks at energy as the using sectors actually receive it, about 53 quads may be called "end-use" energy.[2] If one now relates energy use in the form of heat and liquid fuels to this net or end-use energy, these two categories loom even larger—amounting to some 92 percent of energy as it is actually used in buildings, vehicles, and industry. The other 8 percent represents uses that are necessarily electric—electronics, lights, electroplating, or motors. Actually, electricity is used to provide some heat—space heat, hot water, and cooking, for example, so that it provides some 15 percent of end-use energy. Since there are other ways of providing heat, electricity has been able to penetrate markets beyond those for which it is the only choice. For the user it is clean and convenient, and the conversion devices (strip heaters, for example) may be cheaper than fuel conversion devices (furnaces). This market penetration occurred mostly when electricity was cheap and its real price was falling.

Discussions of energy problems frequently turn out to be discussions of electricity. That is surprising, since one would expect more attention to be paid to the other 90 percent of the problem.[3]

Energy Use in an "Energy-Efficient" U.S. Economy

Given this brief look at energy use in the United States as it was in 1980, it is time to ask how much less energy might have been used in an "energy-efficient" economy. Energy prices were higher in 1980 than they had been expected to be when much of the energy-using capital stock was put in place. Moreover, if 1980 prices truly reflected long-run marginal supply prices for oil, gas, coal, and electricity, energy prices would have been still higher. These observations lead to a definition of energy efficiency as it is used in this study.

The 1980 economy would have been energy-efficient if all conservation-efficiency investments that were at least as profitable as investments in new energy supply already had been made. Put slightly differently, any conservation-efficiency investment that saved a unit of energy at a cost less than the marginal supply price of a unit of energy already would have been made in a conservation-efficiency economy.[4] An example might be useful. In 1980 buyers of refrigerators could purchase more expensive but energy-efficient models. At a marginal supply price for electricity of 10¢ per kilowatt hour, such an investment would yield a positive return of about 45 percent annually. Such returns (tax-free, at that) are rarely available anywhere. Viewed as a way of buying "saved" electricity, the $200 extra investment, at a 6 percent real rate of interst, is "buying" electricity at 1.2¢ per kilowatt hour. Clearly, the 1980 stock of refrigerators was not energy-efficient.[5]

What, then, would have been the quantity of energy used in an energy-efficient United States in 1980? My estimate, supported in detail in the rest of this chapter, is 20–25 quads, some one-third to one-fourth of the amount actually used. That is, the United States could have enjoyed essentially the same output of goods and services, and the same level of comfort, but with much less energy used. This estimate assumes that only existing technology would be used, but that all structures, equipment, appliances, vehicles, and industrial processes would have been selected on the basis of cost-effectiveness against 1980 marginal supply prices for energy inputs. These supply prices were much higher than they had been for some decades and were still not fully reflected in costs paid by energy users. (Estimates for these supply prices follow in a moment.)

As a kind of thought experiment, suppose that all energy users actually confronted these supply prices and that they had correctly anticipated them for a long time. Then the United States would have been energy-efficient in 1980, and energy use would have been in the 20–25 quad range. Or, put another way, if everything could be changed overnight to these efficiency levels, this relatively low energy input would have sufficed. Such things cannot, of course, be changed overnight. Given enough time they will be changed, or else a great body of economic analysis and evidence must be discarded.

This large potential reduction in energy use—some 51–56 quads—exists only because most of the economy's structures and equipment were put in place, and operating styles developed, when energy prices were quite low and were falling in real terms. All of that quickly became obsolete in the 1970s. Fortunately, there is great flexibility in the way in which any particular good or service is provided. Higher prices induce input combinations (and technological change) that use less energy for a given set of outputs. This is hardly news, certainly not to economists. What is perhaps news, and certainly one of the valuable insights into energy policy provided by the last decade, is that the potential is so large.

There is a short-run conceptual difficulty with this approach of comparing investment returns in energy conservation and in new energy supplies. If by 1980 U.S. energy users had invested enough in energy efficiency so as to be using 20–25 quads, then long-run marginal supply prices would have been lower than those given. Producers of oil, gas, and electricity would have "moved down their supply curves." As economists view things, energy users would have overinvested in energy conservation.

A more detailed formulation of the issues would be explicitly dynamic and would trace the time path of energy prices that would maintain similar returns to investment in energy supply and in energy conservation. Since short-run supplies and demands are much less elastic than those in the long run, there would be adjustment lags. The presence of such lags nearly always leads to price fluctuations. These matters are considered more fully in chapters 8 and 9.

In the meantime the simpler formulation—evaluating efficiency investments against 1980 long-run supply prices—is kept. New investments *are* being made in conventional energy supply, while many productive investments in conservation await discovery. The simple approach has the merit of highlighting the large size of the energy conservation potential calculated in that way.[6] The analysis that follows estimates the conservation-efficiency potential with respect to 1980 long-run marginal supply prices for oil, gas, coal, and electricity. It is time now to give estimates for those prices.

Long-Run Marginal Supply Prices

The concept of long-run marginal supply prices, or long-run marginal costs, seems straightforward enough. One asks what price must be offered to secure additional units of the good in question. For petroleum (and for natural gas) the calculation is somewhat elusive. Does it mean, for example, the price for oil would have to be paid in 1980 to get additions to proved reserves in the amount of some 6 billion barrels (1980 U.S. consumption)? If so, then the supply price would ascend through time as the most easily exploitable reserves are used. Does it mean oil in the United States, or anywhere in the world? At what risk premium for foreign oil? Does it mean the price at which the first available substitute comes to the market: liquid fuel from coal or oil shale, for example? How do we treat the great overhang on world oil markets that results from curtailed use and "excess" reserves?

Estimates are, therefore, somewhat arbitrary. The estimate used here is both arbitrary and conservative. Should the reader prefer some other, especially higher, marginal supply price, an extra degree of conservatism will have been added to my estimates of energy saving.

For oil, the arbitrary choice per barrel of oil is $34 in 1980 prices, or about $40 in 1985 prices. This was the price attained (except for briefly higher prices in the spot market) after the "second oil crisis." When the price of oil was this high and was expected to rise, there was enough exploration activity to maintain oil reserves. Those reserves were about as high in 1983 as they were in 1973. Oil used up during the decade was replaced in reserves.[7] This is what "supply price" means—a price sufficiently high to call forth a given quantity of new oil. The quantity, about 20 billion barrels of new reserves, is simply the quantity used up each year that must therefore be replaced. Short-run prices may be much lower, as they were into mid-1986. This means that for a while reserves are being used up faster than they are being replaced.

The decline in real oil prices since 1982 offers evidence of the many opportunities to conserve oil rather than evidence of "excessive" new supplies. Of particular concern to Americans is the decline in domestic reserves since 1973. If the world long-run marginal supply price for oil is $34 (1980 dollars), then the domestic long-run supply price must be still higher. The first U.S. plant to produce oil from shale is still not operating properly, and it has a price guarantee set at $42.50 per barrel (in current dollars). Still higher, a World Bank study[8] uses $56 per barrel (1981 dollars) as a price that will trigger large outputs of petroleum substitutes. On the lower side, another World Bank study estimates that in the largest oil-importing developing nations, new oil could be added to reserves at $10 to $20 a barrel in the most favorable cases, and nearly everywhere at $25 per barrel.[9] The quantities generally are smaller than those used in the United States in 1980 (or even by an energy-efficient United States in the 1980s). These new supplies, then, would not reach the margin nor change the figure arbitrarily chosen—$34 per barrel.[10]

A much lower long-run figure has been used by Morris Adelman. He observes that the oil resources most readily developed into reserves lie in the Persian Gulf area. These were not developed extensively during the past decade since production there was below capacity most of the time. The two oil shocks and the price increases that resulted stimulated exploration in the rest of the world where costs are higher. That would explain the pattern I have observed—that a price of $34 did not expand world reserves but merely maintained them.

This is a reasonable argument for a lower price. I stay with the higher figure for several reasons. It recognizes the political-strategic forces that enter into oil price determinations. It includes a "security premium" for the United States. Since in each passing decade the world now uses one-twelfth to one-tenth of all the oil available to the human race, it will not be long before $34 is an appropriate estimate even in the lowest-cost areas of the world.

The reader is again reminded that short-run prices may be much lower than my figure without destroying its validity as a long-run guide to energy policy. Short-run prices need only cover the cost of operating wells that have already been developed. These could be as low as $10 a barrel, or even less during the period when some producers are being forced to shut down.

In order to make comparisons among fuels, analysts frequently compare their heat-equivalent contents. Since a barrel of oil provides about 5.8 million btu's, oil at $34 per barrel implies a cost per million btu's of a little less than $6.00. This figure is for crude oil. Refined products would include processing, transportation, and dealer markups. Gasoline before taxes would be about $10 to $12 per million btu's at the indicated price for crude oil, and home heating oil about $8.00.

For natural gas, an upper limit is provided by oil prices. Enough users have the capability to make quick conversions (utility boilers and industrial users) so that price divergences quickly lead to changes in consumption. Partial decontrol in the United States, and the presumably expected higher future prices, stimulated new discoveries and net additions to reserves. It is the situation on the North American continent that is relevant to the United States. Natural gas in many areas remote from markets in industrial nations is much cheaper but is costly to transport. The estimate used here is a wellhead price of $3.50 to $4.00 per thousand cubic feet, or delivered prices to consumers at $6 to $8, for a long-run marginal supply price.[11] Gas-btu conversion is easy to remember: the above figures for volumes of gas also are approximately correct for prices per million btu's.

For electricity, it appears that plants started in 1980 will result in delivered electricity prices to the customer in the range of 10¢ to 12¢ per kilowatt hour. Some nuclear plants now being finished in the United States will generate electricity costing much more than this. That, of course, is why no one is ordering any more nuclear facilities. The figure used here is based on a new coal plant with scrubbers. The reader may have seen lower figures elsewhere. The above 10–12¢ includes transmission and distribution costs, franchise fees, and anything else actually paid by purchasers of electricity. These are "first-year" prices—not "levelized" prices where costs escalate with inflation but are discounted back, both at arbitrarily selected rates. Buying electricity at 12¢ per kilowatt hour, incidentally, is like buying oil, in heat equivalent terms, at $203 a barrel.[12] This is an important point; it explains succinctly why electricity is not likely to make greater inroads in energy markets where heat is sought, especially in industry.

Coal costs, averaged across the United States, were $1.35 per million btu's in 1980. Transportation costs vary regionally. Coal is clearly the cheapest fossil fuel in terms of dollars per million btu's. Yet relatively little coal is used except in generating electricity and in making steel. In 1980 these two uses accounted for 81 percent and 10 percent of coal use, respectively. Other industrial uses amounted to 9 percent. Coal use is very small in the residential, commercial, and transportation sectors. In industry, oil and gas continue to be the fossil fuels of choice; wood use is as large as coal use.

Since coal is relatively cheap, one might ask why there has not been much more conversion to coal in industry. This conversion, incidentally, was expected to be one of the main factors in reducing oil use. The answer is that coal requires expensive equipment for handling and for clean combustion.

The long-run marginal supply price of coal does not rise very much in real terms with expanded output. Coal reserves are very large, and, given time, the industry can expand without much cost increase. But that cost is not relevant to the 80 to 85 percent of coal that is used in producing electricity. Rather, it appears as the long-run marginal supply price of electricity to energy buyers—a topic already covered.

For coal-based synthetic fuels it is the cost of conversion that dominates, not the cost of coal. We have no figures based on operating experience; the lowest estimates run from $40 to $45 per barrel.[13]

New technologies that are entering the market burn coal more efficiently and with little pollution. These are the most likely to increase coal use in industry when costs become competitive.[14] The long-run marginal supply price of coal, then, is quite low but not particularly relevant.

Far more relevant is the fact that coal combustion creates more carbon dioxide per unit of energy than does the combustion of oil and gas. There has been increasing concern about the buildup of CO_2 in the atmosphere and the resulting change in climate. These costs (discussed in more detail in chapter 9) could be so large as to make the costs of burning coal prohibitively high.

Long-run marginal supply prices for synthetic fuels also should be mentioned again. Estimates for coal-based fuels were just given. Other possible sources are oil shales and tar sands. The lowest figure given for oil from shale is the $42.50 per barrel cited for Unocal's plant. That is a federal price guarantee that will, apparently, not be profitable even if the plant can be made to work. Canadian tar sands have provided small amounts of crude oil for some years; as new facilities are built, it appears that costs will be in the vicinity of the petroleum costs already mentioned.[15] All synthetic fuels, then, appear to have long-run marginal supply prices higher than those given for petroleum, or, at best, no lower than those prices. Any conservation investment that is cost-effective with respect to petroleum also is cost-effective with respect to synthetic fuels.

The 1980 long-run marginal supply price for crude oil is about eight times that of pre-embargo 1973 prices in real terms. For delivered oil products, gas, and electricity, the marginal supply prices are, respectively, about 2, 3.5, and 2.5 times 1973 real prices. Some demand adjustments already had taken place by 1980, but more remain to be observed. Lags can be quite long, and much of the price increase took place late in the period. We already have asserted that, for prices as high as those just given, an energy-efficient United States in 1980 would have used 20–25 quads, not 76. The next section looks in detail at the derivation of that estimate.

Potential Energy Gains by Sector

This section is intended to establish the plausibility of the threefold to fourfold gain for the United States just mentioned. We examine, in turn, energy use in buildings, in transportation, and in industry, following now well-established conventions in energy-use analysis.

Energy efficiency in buildings.   Residential and commercial structures now exist that require virtually no conventional energy for space heat (which is the largest end-use category in the buildings sector). They are now found in climates that range from the very cold (Massachusetts or Canada) to mild (Florida, California). Modern comfort standards found in centrally heated buildings are maintained in these new structures by solar gain and by heat from lights, appliances, and people. Costs in these new structures for energy-conserving features are well below long-run marginal supply prices for new energy supplies.[16]

These results are obtained mainly by the extension of well-known techniques for reducing heat loss: insulation in attics, walls, foundations, and slab edges. Very low infiltration rates and window treatments do the rest. Ventilators with air-to-air heat exchangers avoid indoor air pollution. The extra cost of thermal integrity and passive solar features is partly offset by the downsizing of heating systems (or indeed their elimination). The rest is quickly amortized by fuel or electricity savings. These living examples lend support to a Los Alamos study that concludes: "the potential exists for passive solar supply of almost all space heating needs in the U.S."[17] A study by the Solar Energy Research Institute also concluded that over 90 percent of space heating needs can be so met. I therefore use 90 to 95 percent as an estimate of the long-run potential reduction in energy use for space heat.

It is easy to find well-insulated or passive solar houses that do not perform at this level. It takes quite a bit of tinkering and design experimentation to reach this level of efficiency in any given climate. *Some* such structures exist in sufficiently diverse climate zones so that, given learning time for designers and builders, they can be built anywhere in the United States.

Similar design techniques plus a few others can considerably reduce energy use for space cooling. Conventional cooling systems can be eliminated by good design in all climates in which the twenty-four-hour average summer temperatures and humidities are in the comfort range. (That range can be extended some 6–8 degrees Fahrenheit if people do not mind using electric fans.) Passively cooled structures also exist in hot, dry, summer climates where clear night skies permit heat dissipation by radiation (as in the U.S. Southwest, for example). In the remainder of the United States (especially the southeast) presently available techniques cannot maintain modern comfort standards without cooling/dehumidifying devices.[18] Good design, however, can greatly reduce cooling loads[19] (chiefly by reducing heat gain from the sun and, especially in commercial buildings, using fewer and more efficient lights) and then can meet this reduced cooling load with much more efficient equipment. The most efficient air-conditioning units now on the market are about twice as efficient as the average of those now in place.[20] I take a saving here of 75 percent in the long run. This may be thought of as roughly a 50 percent reduction in cooling loads, with the remaining loads met by equipment that is twice as efficient.[21] To be sure, improvements in the thermal integrity of buildings reduce heating loads more than they do cooling loads, since temperature differentials inside-to-outside are greater in the winter. A 90 percent heating load reduction might correspond to, say, a 30 percent cooling load reduction. The rest of my 50 percent reduction is accomplished by more efficient appliances (discussed below) that emit less heat, by more efficient lighting, by design that avoids running chillers when outdoor air will do, and by reduced solar gain.

Commercial buildings require somewhat different conservation strategies in space conditioning than do residences. In the 1960s electricity was relatively cheap, declining in price, and widely expected to be the cheap energy source of the future. Space-conditioning systems were quite electricity-intensive in new commercial buildings. Chillers ran year-round, for large commercial buildings frequently needed cooling in their cores even in very cold weather. Chilled air was then reheated by electrical resistance heaters to reach the desired temperature. This kind of system design also was used to reduce humidity in areas where that is a problem. This design may have seemed appropriate with electricity at 1¢ per kilowatt hour and the prospect of lower prices in the future. It seems much too energy-intensive under present conditions. Systems that maintain indoor comfort with much less electricity are being designed and installed.[22]

Lights use about one-seventh of the nation's electricity. Such demands can be reduced by increased daylighting, further reductions in overlighting (some reduction had already been achieved by 1980), and not lighting unoccupied rooms. The reduced lighting demand can then by met with more efficient bulbs, and with automatic controls where lighting auxiliary needs fluctuate with daylight levels. This combination of actions can save 80 percent of lighting electricity. Efficiency gains of 80 percent in bulbs alone are available as new products replace today's incandescent bulbs.[23] Electronic ballasts raise the efficiency of fluorescent bulbs about

40 percent. These bulbs are already much more efficient than incandescent bulbs. For fluorescent lights the other savings come from daylighting and controls. Costs of these measures have been estimated at 0–2¢ per kilowatt hour[24]—they would have been good investments for many years. The new efficient bulbs are admittedly more expensive. They last longer, however, so that lifetime operating costs are in the range indicated.

Among appliances, fourfold gains in refrigerator efficiency have been achieved. Unfortunately for us, that is in Japan;[25] the best American models now on the market are about twice as efficient as the existing average. American appliance makers adamantly oppose appliance efficiency standards,[26] just as their Detroit counterparts opposed auto mileage (efficiency!) standards a decade ago. Japanese appliances are now entering U.S. markets. I take 75 percent as the long-term gain for refrigerators, though larger gains have been made in prototypes. For other appliances I use 50 to 70 percent. For stoves, examples of energy-reducing actions include better insulation, replacement of pilot lights with electronic ignition, new combustion technologies, better conduction from burners to pans,[27] and the further penetration of microwaves. Clothes dryers can have heat recuperators added.

Hot water usage can be reduced by such devices as shower head flow restricters and water-economizing washing machines and dishwashers. The attendant saving of water is not of concern here but is important in other contexts. Water tank insulation and timers save energy in heating the reduced amount of water. In cold climates where the expense is warranted, heat exchangers can preheat incoming water from outgoing waste water. For those who do not get solar water heaters (discussed in Chapter 3 under renewable energy technologies) waste heat from air conditioners or refrigerators can be used. Simply lowering the temperature where it is set higher than necessary should already have been done, but has not been done everywhere. I use a saving of 60 percent here.[28]

The weighted average of all of the above reductions is about 80 percent.[29] That is, America's buildings could be as comfortable and convenient as they are now but consume 20 percent of the energy that they did in 1980. An independent confirmation of this estimate comes from the commercial sector. The Solar Energy Research Institute (hereafter SERI) projects reduction in commercial building energy use of 85 percent with advanced designs. The California Energy Commission projects 80 percent reductions.[30] Reductions of 77 to 80 percent have already been achieved in New Jersey, Colorado, and North Carolina.[31] All of these are at costs lower than the long-run marginal supply prices of oil, gas, or electricity.

Transportation.   In the transportation sector about 45 percent of the energy used goes to automobiles. The rest is used in trucks (including light trucks), buses, aircraft, trains, freighters, barges, and pipelines. U.S. auto gas mileage reached its lowest point in 1973—13 miles per gallon. By 1980 the average for the entire fleet had risen to 15, and by 1983, 16.7.[32] This figure will probably reach 25–26, realized on the road, not calculated from tests, as the entire auto fleet turns over. New policy initiatives could raise the figure considerably, but the issue in the United States at the moment is the maintenance of the 27.5 standard mandated in 1975 legislation. Imports have higher averages, so that the 25–26 figure given seems to be attainable. Although this is far short of the mileage that is technically attainable and economically warranted, it doubles 1973 mileage and raises the 1980 average by 67 to 73 percent. Since about half of the American auto-buying public apparently has neither foresight nor even hindsight, an oil crisis with gas lines about every six years may be a regrettable necessity for rapid change.

SERI makes the case for a potential fleet average of 75 mpg, a figure which should not be too readily dismissed as a flight of fancy. British Leyland built a five-passenger prototype in 1982 that averaged 63 mpg; 133 (one hundred and thirty-three) at steady 30 miles-per-hour driving.[33] Several models on the market in 1985 got 50 miles per gallon or more; one reached 60 on the highway.[34] General Motors has built a prototype minicar with mileages of 95 and 68. It has no plans to market such a vehicle, but similar minicars have begun to penetrate markets abroad. General Motors is also introducing a U.S. minicar with mileages of 53 and 68.[35] Its "Saturn Project" vehicle is expected to achieve 45 mpg in city driving.[36] A Renault prototype is reported to have reached 94 mpg.[37] Battelle Memorial Institute concluded that, by combining only existing efficiency-raising techniques into one vehicle, a four-passenger, 100 mpg vehicle could be built.[38]

Research officials at Ford, writing in *Science*, examined trends in auto technology. They speculate as follows about the "average" vehicle of the late 1990s: "a four or five passenger vehicle in the 2,000-pound inertial weight class—and fuel economy in excess of 100 miles per gallon on the highway—the appearance of such a vehicle will depend strongly on the price of fuel."[39] Much of this weight reduction is with lighter materials, not just downsizing. Among these lighter materials are plastics and aluminum. The initial manufacture of aluminum requires more energy than steel, but that is quickly made up in fuel savings, and made up again when aluminum is recycled at one-sixth the energy cost of recycled steel.

The many studies that find high-mileage autos "cost-effective" (e.g., Ross and Williams, Gibbons, SERI) do so by comparing retooling and production costs in the auto industry with the value of fuel saved. On that basis, following their practice, I have used a five-fold gain in auto fuel efficiency in my calculations. As an exchange of letters in *Science* points out, European and Japanese gas prices are about twice as high as those in the United States, and automobiles there average about 30 mpg.[40] We shall probably not, then, see prices in this century that would raise fleet averages to 50–75 mpg if matters are left entirely to the auto manufacturers and the vehicles they choose to put before the public. Such prices will certainly be seen eventually. We simply do not know what the "equilibrium" fleet average mpg vs. liquid fuel cost is. In any event, the technical potential for high-mileage vehicles is there when it is needed, and the changeover is "economic" for auto manufacturers at 1980 long-run marginal supply prices.

For air transport, new models like Boeing's 757 and 767 are 35 to 40 percent more fuel-efficient per seat mile than earlier aircraft.[41] This gain comes after a 33 percent gain already achieved by the airlines in the 1973–81 period.[42] The next generation of aircraft, already well along in design, promises to add another 15–20 percentage points. Therefore, I use 50 percent for air transportation.[43]

Efficiency gains in truck freight are steadily being achieved with air deflectors and better maintenance. Fruehauf reportedly has a prototype with gains of 40 percent over existing vehicles.[44] More innovations are under way in engine efficiency, improved tires, and better transmissions. Deregulation has reduced the number of empty return trips. A small shift back to more fuel-efficient rail freight is occurring as a fuel cost differential becomes more visible. SERI takes a 50 percent overall saving on rail and truck freight. Since this is a twenty-two-year partial adjustment, and I am looking at technical and economic limits, I take 60 percent as a conservative result based on the same data.[45] Rail efficiency gains are also taking place, though most other studies do not consider such gains.[46]

In buses and subways energy now dissipated as heat from brakes can be partially recaptured. Some buses with flywheels and others with pressure tubes now operate in Stockholm and in Copenhagen. Savings of 25 percent in fuel use are reported.[47] Captured energy from braking, temporarily stored, is used to assist the next start-up. Subway electric motor drives can be designed to run backward as generators, thus feeding electricity into the lines during braking. This saving can be quite important in some areas. About 7 percent of New York's electricity runs the subway and PATH train system. Regenerative braking (along with more efficient lights) could reduce this use by one-third.[48] Thus, the equivalent of a 100-megawatt power plant sits unused in subways.

The overall saving in the transportation sector exceeds two-thirds; the details are shown in the appendix at the end of this chapter.

Industry.   This is the third major sector for recording energy use in official statistics. It includes agriculture, mining, and construction as well as manufacturing. Even within manufacturing, it is an extremely heterogeneous sector, though energy use is highly concentrated. Six industry groups (metals, paper, petroleum refining, chemicals, stone, clay and glass products, and food processing) account for 80 percent or more of manufacturing energy use. Industry apparently responds rather quickly to fuel price changes (or prospective changes). This sector accounted for perhaps two-thirds of all U.S. efficiency gains from 1973 to 1979.

Industry used 30 quads of energy in 1980. Of this amount, about two-thirds went to the two main uses: process heat (12.5 quads) and mechanical drive (7.8 quads). Both areas present abundant opportunities to use energy more efficiently. Primary fuels for process heat (little electricity is so used as it is too expensive) can have their usage considerably reduced. Furnace insulation, combustion controls, heat recovery, combustion of stack gases, continuous processing of heated materials—these techniques are widely applicable in industry. Electric motors for mechanical drive provide major opportunities for efficiency gains through power-factor controllers and variable speed drives.

SERI projects an overall efficiency gain of 33 percent by the year 2000 if cogeneration of electricity and heat is included. (The CONAES study[49] for the National Academy of Sciences projects a 40 percent gain by the year 2010.) This study projects a long-run gain of 50–60 percent, a figure that will probably be surpassed. As we shall see in chapter 6, calculations for Sweden and West Germany, already more efficient than the United States, show larger gains. Still larger gains are shown for energy-inefficient Britain.

One must be careful here. The studies cited all assume continued economic growth, with a falling industrial share (though industry grows in absolute terms) and a still larger fall in the share for energy-intensive materials output. My estimate, on the other hand, refers to an alternative way of producing the 1980 Gross National Product and cannot, therefore, reap any benefit from a relatively smaller industrial sector sometime in the future. I take the *ceteris paribus* savings in the above studies, extrapolate them out to a full adjustment to changed energy prices, and take some additional savings from reduced materials wastage, recycling, and increased durability warranted by higher relative prices for energy-intensive goods. In order to make the estimate credible, but to spare all readers some tedium, a brief appendix to this chapter includes an analysis of seven especially energy-intensive industries: steel, aluminum, paper, cement, glass, chemicals, and petroleum refining.

These savings of 50–60 percent in industry would be achieved in thirty to thirty-five years if only annual rates of gain in energy productivity of the last decade can be maintained.[50]

Summary of Energy-Efficiency Gains

It is now time to summarize the discussion of the scope for increasing the efficiency of energy use in the United States. We have shown that, in 1980, energy use in residential and commercial buildings could have been about one-fifth of that actually recorded without loss of comfort or with any other loss of amenities. Conservation in this sense most assuredly does not mean freezing in the winter or roasting in the summer.

Transportation energy use could drop by 70 percent with the same number of passenger miles traveled and ton-miles of goods moved. There would be an amenity change with respect to automobiles. They would be lighter—partly by being smaller and partly by using lighter materials. This would be the only change noticeable to drivers and passengers. One other change in the movement of goods should be noted: there would be much less fuel to haul around since two-thirds less would be used. This is not a trivial item; some 40 percent, by weight, of all materials used in the United States are fuels.[51]

A similar qualification must be made for industry. We have, until now, talked of producing the 1980 Gross National Product with much less energy. That is not strictly correct, since a portion of that product had to do with value added in the energy sector. This portion of output would, of course, be reduced. One similar reduction in the industrial sector is made—we allow for the somewhat smaller output of steel that accompanies the weight loss of automobiles. There would be many other changes, especially when second- and third-order effects are taken into account. We make no attempt to correct for these changes; the results are approximate and sufficient to the present purposes.

With these qualifications, then, we assert that the industrial output of 1980 could have been produced with about 40 percent of the energy actually used.

If we sum the savings just discussed in the three sectors, we get approximately the three- to fourfold gain in energy efficiency mentioned earlier. This calculation is shown in the appendix that follows. This is a very large gain indeed, so large that it may be difficult for the reader to accept even after detailed evidence has been examined.

Figures 1, 2, and 3 may help readers to visualize the kinds of gains that are technically and economically feasible. One example is selected from each of the three energy-consuming sectors—space heat, auto fuel, and energy required to produce a ton of cement. The first two uses are now dominant energy uses in their respective sectors. The last, cement, is chosen because it typifies energy-intensive heavy industry and shows a reduction similar to that calculated for industry as a whole.

A study conducted for the American Physical Society and published in 1975 is consistent with the proposition that these findings are quite modest.[52] The economy was then running at a second-law efficiency of 10 to 15 percent. Even that estimate understates the truly vast potential for increasing energy productivity. Why, for example, should one use a highly efficient system to heat a poorly insulated house? The Physical Society estimate dealt with issues like heating system efficiency; it did not consider the thermal integrity of structures.

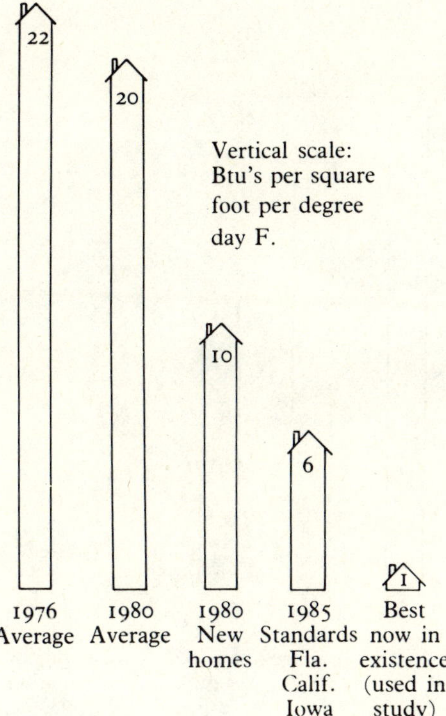

Figure 1   Energy to Heat a Single-Family U. S. Residence (adjusted for size of home and severity of climate). Sources: SERI, *Solar Age* (various issues), California Energy Commission, Florida Department of Community Affairs.

Results in Other Studies

It is now time to compare these results with those of numerous other studies. Many recent studies have shown the presence of large potential reductions in energy use. One was done for the National Academy of Sciences,[53] one by John P. Gibbons (director of the Office of Technology Assessment) and William U. Chandler,[54] one by Roger Sant of the Nixon-Ford Federal Energy Administration,[55] one by Marc Ross and Robert Williams[56] of the University of Michigan and Princeton University, respectively, one by the Solar Energy Research Institute,[57] and one

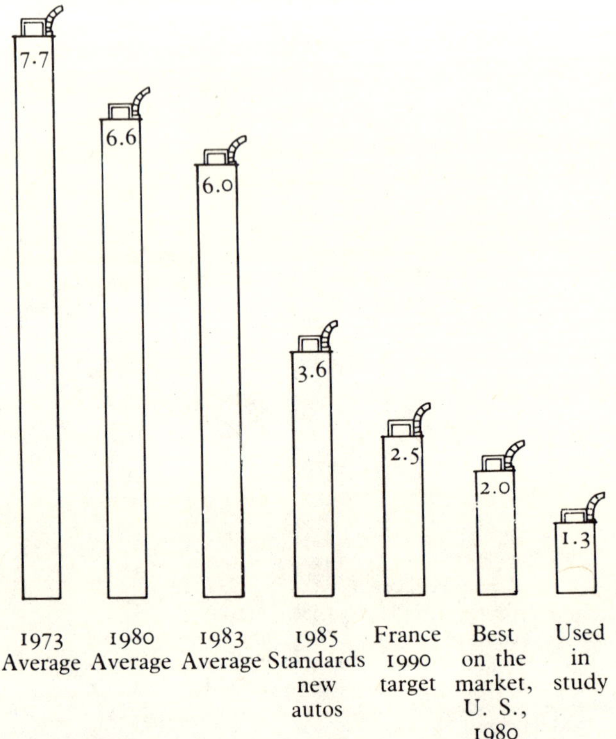

Figure 2   United States Automobile Mileage (gallons of gasoline needed to go 100 miles).

that has received the most attention, *Energy Future*,[58] by a faculty group at the Harvard Business School. A still more recent study comes from the National Audubon Society.[59] This second version of its *Audubon Energy Plan* is quite detailed, is capable of estimating, with modeling, the effects of numerous policy measures, and is scaled to match models used by the Department of Energy and some of the energy supply industries.

Most of these studies imply roughly a halving of the energy-GNP ratio by the year 2000. SERI, for example, has real output rising by 80 percent over the 1977–2000 period, with primary energy inputs falling to about 65 quads—a 53 percent drop in the energy-GNP ratio. This report was submitted in draft form to senior officials in the Department of Energy during the Carter administration. The conservation potential was apparently too large for a department at that time preoccupied with a vast synfuels program. The report was not released by the department; it was, not surprisingly, suppressed by the Reagan administration, and then ignored after it was made public through the medium of congressional hearings.

The SERI conclusions are not really so radical. They are roughly consistent with one of the scenarios in the earlier report prepared by the National Academy of Sciences group. Ross and Williams show still larger gains; Sant shows somewhat smaller ones,[60] as does the Audubon Society.[61]

But how can I assert a three- to fourfold efficiency gain when the studies just cited suggest a twofold gain? The difference is mainly in the way in which the question is framed, not in the underlying estimates of the long-run potential efficiency gain. All of the studies mentioned start at a date in the 1970s or 1980s and then move forward to the year 2000 or

Figure 3 Energy Used to Make a Ton of Cement (million btu's). Sources: *Minerals Yearbook*, World Bank, Governors Energy Office, State of Florida.

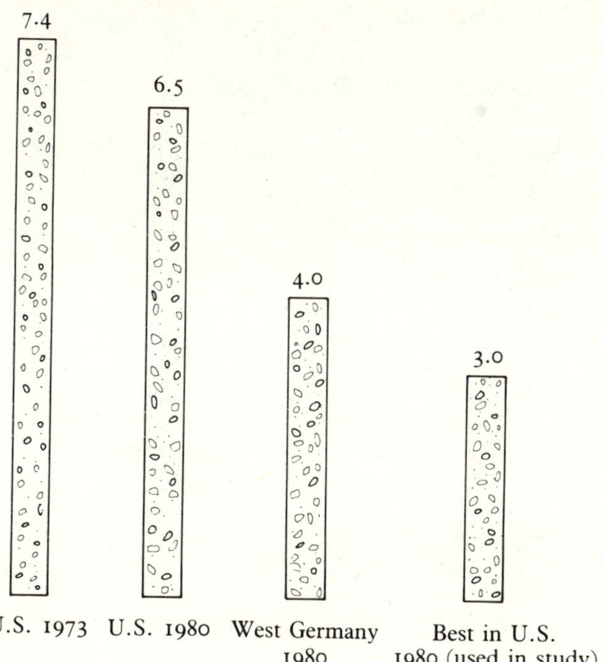

2010. Efficiency-enhancing measures are gradually introduced, while economic growth proceeds. Even at those distant dates only a partial adjustment to more expensive energy has taken place. There are still many energy-obsolete buildings as well as some industrial plants and equipment. A truly long-run adjustment would entail still lower energy-GNP ratios in each of these studies.[62]

Recent Trends in Energy Use

These conclusions are also supported by the experience of recent years. The U.S. economy is moving in the direction of lower energy use per unit of output at surprising speed. Energy use was 1 percent lower in 1985 than it was in 1973, while the GNP had risen by 32 percent. The energy-GNP ratio fell at an annual average rate of 2.5 percent. Figure 4 shows these patterns.

To avoid cyclical distortions, we can compare 1973 and 1979, two cyclical peak years. Over that six-year period the energy-GNP ratio fell 10

Figure 4 Gross National Product and Energy Use, 1973–85, United States. Sources: *Monthly Energy Review*, June 1985 and February 1986. GNP data for 1985 restated in 1972 dollars by author.

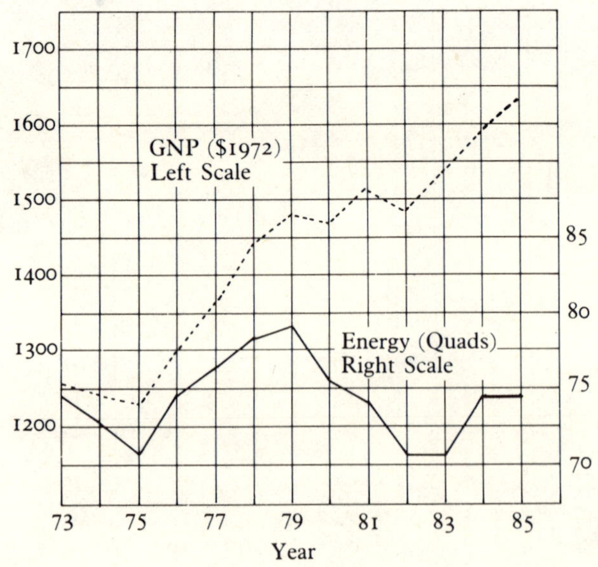

percent. Oil and gas prices were controlled at levels far below marginal supply prices, and electricity prices were also well below long-run marginal costs. Much of this gain resulted from rapid adjustments in the industrial sector. Efficiency gains have apparently accelerated since 1979.[63] Essentially the same thing is going on throughout the world. Europe and Japan, already with energy-GNP ratios lower than that of the United States (by factors of .5–.7)[64] have reduced these ratios still further. Gains over the 1973–80 period range from 11 percent (France) to 22 percent (Japan).[65]

Energy Use Reduction with Conventional Elasticities

The estimates of energy use reduction attending higher long-run marginal supply prices should also be compared with estimates that use conventionally estimated price elasticities. That comparison is now made.

The drop in energy use relative to GNP between 1973 and 1980 is quite consistent with conventional demand theory. Consider the residential-commercial sector as an example. Energy use in buildings was 24.1 quads in 1973. By 1980 real energy prices to households and commercial sector businesses had risen about 50 percent,[66] while GNP (here a surrogate for income) rose about 18 percent. For price and income elasticities we shall use the following, which are taken from an extensive review of the literature:[67] short-term price, −.16; long-term price, −.80; and income, .85. Although seven years is longer than most people's definition of "short run," we shall first use the short-term price elasticity estimate. Price controls and other lags slowed the effects on energy use, and investments in energy efficiency already described were not yet a large factor in reducing energy use.

If we use the above parameters to estimate 1980 energy use, we get 25.5 quads, an estimate quite close to the actual use of 25.7 quads. If the short-run response was so consistent with familiar elasticities, what would happen if we applied a long-term price elasticity (−.80) and marginal supply prices? The latter, incidentally, averaged for energy sources used, were about 2.5 times greater than real 1973 prices, well beyond the 1.5-fold increase in prices actually paid.

This calculation yields a result of 13.4 quads. That is, if households and commercial enterprises had fully adjusted in 1980 to price changes, and if those price changes reflected supply prices as we estimate them, then energy use in buildings would have been 13.4 quads, not the 25.7 quads actually used.[68] No one in government, or in the energy supply industries, seems to contemplate a drop in energy use of this magnitude. Similar results hold for transportation and, less dramatically, in industry.

My own results show a much larger decrease in energy usage. Households, for example, would have used about 5 quads if they were fully energy-efficient. I have used "engineering-economic" sources for each of the major energy end-uses. Implicit long-run price elasticities are larger in part because sharply rising prices induce some technological change. This change is excluded in principle, though only partly in practice, in defining and measuring price elasticities. Then, too, empirically derived elasticities rarely have a data set that includes price changes of these magnitudes. The world may operate according to stable, log-linear functions over very wide ranges, but there is no reason to suppose that it does. Time will tell who is right.

Indeed, some investigations, including those referred to earlier in this chapter, show that many efficiency investments were already warranted at 1973 prices but had not been undertaken. Time and effort are involved in identifying these opportunities. That time and effort are more likely to be expended when energy prices are rising than when they are falling.

In any event, many profitable energy-saving investment opportunities have been identified throughout the economy. If some activity is possible and profitable, a market economy will sooner or later find it out. Later sometimes is much later; hence, one case for energy policy interventions.

Chapter Summary

This chapter has dealt with a wide range of material and contains many important conclusions. A summary may be helpful.

The United States used 76 quads of energy in 1980, the base year for these estimates of the U.S. energy conservation potential. If one estimates 1980 long-run marginal supply prices for oil, gas, coal, and electricity, he can then calculate how much energy-conserving investment would have been economically superior to new energy supply investments.

An examination of the way in which energy is used in the building, transportation, and industrial sectors indicates that an energy-efficient

America would, in 1980, have consumed 20–25 quads, a decrease of 67 to 75 percent. Many other studies suggest decreases of 40 to 50 percent by the year 2000. These estimates are not necessarily inconsistent with mine. They involve partial adjustments to prices seen by energy buyers; mine are long-run adjustments to (higher) long-run supply prices. Price controls, subsidies, institutions for pricing electricity, and other factors kept proper price signals from reaching energy buyers.

My results also go considerably beyond those derived by using most coefficients of price elasticity from the literature. The reasons for that difference are explored. Even those elasticities suggest much larger decreases in energy consumption than any contemplated in government or in the energy supply industries.

Energy use patterns in the 1973–85 period clearly confirm the existence of large potential gains in energy efficiency.

Appendix

Summary of Conservation-Efficiency Gains

The narrative in chapter 2 has described numerous ways in which energy efficiency may be increased in the economy. Numerical estimates of potential reductions in energy use have been given for most end-use categories. These estimates must now be brought together into an economy-wide summary.

Table 2 shows 1980 energy use in the United States in the three energy-consuming sectors. It then shows the much-reduced energy use for each sector that would have occurred in 1980 if all of the conservation-efficiency investments mentioned in chapter 2 had already been made.

Table 2   Summary of Potential Efficiency Gains, 1980.

Sector	1980 energy use (quads)	Efficiency gain coefficient	Adjusted energy use (quads)
Household and commercial	25.7	.17–.22	4.4–5.7
Industrial	30.6	.36–.42	11.0–12.9
Transportation	19.7	.285	5.6
Total	76.0	.28–.32	21.0–24.2

The Household and Commercial Sector, for example, used 25.7 quads in 1980, but could have managed as well with 4.4–5.7 quads in really well-insulated buildings with efficient lights, appliances, and equipment. These figures are calculated using the "efficiency gain coefficients" of .17–.22. The figure of 25.7 is multiplied in turn by .17 and .22 to get 4.4 and 5.7 quads. The coefficients are derived from table 3. They assert that energy use in residential and commercial buildings could have been 17 percent to 22 percent of 1980 actual use.

Table 2 is a summary; each line is taken from a more detailed table for each sector. In addition to table 3, which provides supporting information for the Household and Commercial Sector, table 4 deals with transportation and tables 5 and 6 deal with industry.

Chapter 2 asserts that 1980 energy use of 76 quads could have been reduced to 20–25 quads. The lower figure of 20 quads is not quite attained in table 2. Reaching that figure, rather than the 21.0 quads shown in the table, entails some of the following: reductions in electricity transmission losses and more efficient generation,[69] recycling of lubricating oils, recycling of materials in tires, a reduction in energy-intensity of an economy with less fixed capital in the energy industries, and some substitution of communications for travel in response to higher travel costs.

Table 3 is basically the calculation of a weighted average efficiency gain coefficient for households. The text in chapter 2 asserts that space heating energy demands can be reduced by 90–95 percent, yielding an efficiency gain coefficient of 5–10 percent. The weight applied is 46

Table 3   Energy Efficiency Gains in Residences, Calculation of Weighted Average.

End use	Percentage of total	Efficiency gain coefficient	Portion of remaining use
Space heat	46	.05–.1	2.3–4.6
Space cooling	13	.25	3.25
Hot water	9	.4	3.60
Lights	12	.2	2.40
Refrigerators	6	.25	1.50
Other	14	.3–.5	4.2–7.0
	100		17.25–22.4%

Table 4 Energy Efficiency Gains in Transportation: Calculation of Weight Average Gain.

End use	Percentage of total	Efficiency gain coefficient	Mode shift effect	Fuel volume decrease effect	Adjusted efficiency gain coefficient
Autos and light trucks	58	.2	.95	—	11.02
Freight trucks	13	.5	.75		4.88
Airplanes	12	.5	.90		5.40
Water transport	10	.8		.5	4.00
Pipelines	3			.3	.90
Rail	3	.8	1.16	.6	1.67
Bus	1	.6			.60
	100				28.47

1. In addition to the efficiency gains cited in the text of chapter 2, the following reductions are reflected in the above table: 25 percent of truck freight moves back to rails as more fuel-intensive trucking becomes relatively expensive with 1980 fuel supply prices. This is reflected in a further 25 percent drop of trucking fuel (calculated *after* efficiency gains—no double counting) and a corresponding increase in rail fuel.
2. The reduced use of fuel throughout the economy reduces fuel shipments by pipeline, barge, and rail.
3. High-speed passenger trains displace 5 percent of auto trips and 10 percent of air trips.

percent, which is the share of household energy used for space heating. The resulting weighted average for all household energy uses, 17–22 percent, is then taken to table 2.

In the same manner, table 4 calculates a weighted average efficiency gain coefficient for transportation. Autos and light trucks use 58 percent of transportation energy and are subject to energy use reductions of 80 percent. An efficiency gain coefficient of .2 is multiplied by 58 percent. As the note to the table explains, additional adjustments are made for the effects of shifting some passenger and freight transportation to rails and for the diminished transport of fuels.

The result of the calculation in table 4 is a transportation efficiency gain coefficient of 28.47 percent. That figure is carried to table 2, rounded to 28.5 percent, and then applied to the 19.7 quads of energy used in 1980 for transportation.

The calculation for industrial energy efficiency gains is somewhat more complicated. That subject is treated in the last section of this appendix, and the results are brought forward to table 2.

Finally, a word about the implications for electricity use of all of the above.

Electricity is essentially irreplaceable in lighting, refrigeration, space cooling (unless dessicant cooling with solar recharge becomes available), electronics, electroplating, and industrial motors. It also is used now in many "optional" areas such as ranges and clothes dryers, not to mention space heat and hot water. If "optional" uses except for space and water heat are continued, our energy-efficient economy would use about 8 quads (primary equivalent) of electricity. In a 20–25 quad energy total, electricity would account for 33–40 percent of all energy demand.

This is not an unexpected result; some of the largest potential gains are in space heat, auto transport, and industrial heat—areas in which electricity use has been limited by its high cost. It is true that about 50 percent of *new* homes have electrical heat, but this amounts to only about one-sixth of all homes. Most new construction is in the South and West; electric heat has a higher share of the market in the South.

Since the higher gains in energy efficiency are in nonelectric areas, it follows that electricity will loom larger in an energy-efficient economy. People in the electricity supply industry talk hopefully of new industrial markets. We already have shown the immense competitive disadvantage for electricity in providing heat. Nevertheless, there are some uses in which electricity is so much more efficient that it overcomes the high-cost disadvantage. Mini-mills, which heat scrap in electric arc furnaces and make some steel products, have done well. Induction heat has advantages in some applications. Some drying or curing operations for which fuel has been used can be done more cheaply with induction heating or infrared or ultraviolet radiation. These seem to be special cases—the overall market for industrial process heat is beyond the reach of electricity until competing energy sources are much more costly.

Energy Savings in Industry, 1980

This section gives more detailed estimates of potential gains in energy efficiency in American industry. It supports the brief narrative in chapter 2 and the figures in table 2.

Table 5   U.S. Industrial Energy Use, 1978.

Category	Energy use (quads)	Examined in following narrative	Not treated here
Chemicals			
Feedstocks	2.6		2.6
Energy	4.0	4.0	
Subtotal: Chemicals	6.6		
Metals:			
Iron and steel	3.1	3.1	
Aluminum	.9	.9	
Other	1.6		1.6
Subtotal: Metals	5.6	4.0	
Petroleum refining	3.7	3.7	
Paper	2.6	2.6	
Stone-Clay-Glass:			
Cement	.5	.5	
Glass	.5	.5	
Other	.5		.5
Subtotal: Stone-Clay-Glass	1.5		
Food Processing	1.3		1.3
All other manufacturing	4.5		4.5
Subtotal manufacturing	25.8		
Mining and construction	3.3		3.3
Agriculture	1.6		1.6
Total "industry"	30.7	15.3	15.4

Numbers quickly become confusing: they are quoted for "industry" in U.S. and other official statistics, with or without mining, agriculture, construction, forestry, and fisheries. Moreover, purchased electricity is sometimes shown at its heat equivalent at the point of use, and sometimes at the heat content of the fossil fuel used to produce it.

In order to be clear, this analysis is based on the data in table 5. All purchased electricity included in the above figures is measured "gross"— on the basis of the fossil fuels that would be burned to produce it. Or, put slightly differently, electricity generation losses are "imputed" to the industrial sector.

As table 5 indicates, the following paragraphs will deal with potential efficiency gains for about half the sector. Data for 1978 are used as a proxy for 1980; this makes for consistency with SERI and takes advantage of the 1977 census of manufacturers so that the data are better. Use in 1980 was about 3 percent below that for 1978, though industrial production was at about the same level. Some slight distortion arises from a change in the industrial mix; cement and steel production were down somewhat, but other, generally less energy-intensive items were up in 1980 as compared to 1978. These are very small distortions with little effect on the overall findings. The table also includes about 1.5 quads of energy from wastes in the forest products industries, energy not measured in official U.S. statistics.

Long-run potential efficiency gains based on known technologies that were cost-effective against 1980 marginal supply prices for electricity and fuels are used in the following discussions. We are attempting to determine what industrial energy use in 1980 would have been if all of those production techniques and processes had been used in 1980.

Steel.   U.S. production costs are relatively high for steel, due in part to the labor costs so often cited by the industry. Another major but less often discussed cost disadvantage in an industry with much obsolete plant is excessively high energy use. A 55 percent reduction in energy use is available in well-designed new plants with continuous casting, heat recuperation, by-product gas combustion, and cogeneration. Such efficiencies already are embodied in the world's best plants, though still more improvements are on the way. (Some are discussed in the references for chapter 6 dealing with Sweden.)[70] Increased scrap recycling with the use of electric furnaces also reduces overall energy use.[71] If the 1980 autos had been smaller and lighter in a manner consistent with the mileages in the transportation section, steel production also would have been lower by 5 to 10 percent.[72] Increased product durability and increased remanufacturing of steel-using machinery, vehicles, and appliances contributes another 10 to 20 percent. All of these make a 61–67 percent reduction possible. Coefficients of .33–.40 are therefore applied to energy use in steel in table 6 below.

Aluminum.   New processes for aluminum reduction, the most energy-intensive step in the transformation of ore into aluminum products, reduce energy consumption by fully one-third.[73] (Still newer technologies in the pilot stage in Japan use coal instead of electricity and use much less energy. That reduction is not assumed here.) If electricity used in aluminum production in 1980 had been priced at its marginal cost, much additional recycling would have been warranted. Container legislation, which would have been adopted long ago by any society both informed and concerned about its energy future—not to mention landfill space and aesthetics—would contribute to the increased recycling. Without a detailed analysis, I simply raise the 33 percent saving to 50 percent.[74] This is then applied to the .9 quad in table 5, since this is the energy used in reducing alumina to aluminum. (The rest of the energy used in the industry is included in "other.")

Paper.   Dow Chemical in 1975 listed ways in which energy used in the paper industry could be reduced by 33 percent.[75] One of the large items was increased cogeneration (an activity that has indeed been increasing rapidly). Process improvements already known contributed the rest. Swedish studies indicate long-run savings of 54 to 62 percent in an industry already more energy-efficient than is the case in the United States, but with a different product mix. Gains of 50 to 60 percent seem to be conservative.

Cement.   This industry in the United States has been plagued by overcapacity and obsolete equipment, but there is not enough profit or forecasted growth to warrant investment in new, energy-efficient plants. For these reasons, only modest savings are proposed by studies like SERI. This study asks a different question—what would cement industry energy use have been in 1980 if all investment in capacity had correctly foreseen 1980 marginal supply prices. The best estimator I have found is a plant built by Gulf & Western's Marquette Company.[76] It claims to use 60 percent less energy than the industry average. The claim is supported by the experience of the Florida Mining Company. It already has reduced energy use to 60 percent of the industry 1980 average, and will go below 40 percent of the average with cogeneration.[77]

Chemicals.   This diversified industry is the largest single industrial energy user—over 6.5 quads. Of this, 2.6 quads is for feedstocks, mostly oil and natural gas, a figure that can be reduced only by using less of the products, reducing process wastage, or using other materials in the products. While there are possibilities here, we will focus for the moment on energy fuels, not feedstocks. Efficiency gains have in fact been taking place quite rapidly. Union Carbide, for example, reduced its energy use per pound of product 15 percent in six years, lending credibility to its target of 30 percent in thirteen years.[78] Dow Chemical has had a reputation since its founding for efforts to minimize energy use. If any industry can be counted on to achieve steady gains in energy efficiency, it is chemicals.[79] This industry is, incidentally, a prime candidate for cogeneration—an engineering finding increasingly borne out by installations. The category "inorganic chemicals," incidentally, includes uranium enrichment for nuclear plants—a highly energy-intensive process. This .2–.3 quads would simply disappear in a non-nuclear scenario. I take 50 to 60 percent as an estimate of energy-use reductions.

Petroleum refining.   Energy use in this industry is subject to two reductions: gains in efficiency in the processes, and reductions in petroleum use throughout the economy. In order to avoid lengthy documentation, I take 20 percent for the former[80] (probably already surpassed in the best U.S. refineries), and 67 percent for the latter, based on efficiency gains in transportation. These reductions are not additive, but multiplicative, and yield a combined saving of 74 percent.

Glass.   The float process reduces energy use in making flat glass by some 30 percent. Other measures that apply to high-temperature processes in general (heat recuperation, insulation, etc.) add further possibilities. In the case of bottles for beverages and foods (about two-thirds of the industry's output) reuse of containers (not remelting into more glass products) would save 67 to 83 percent of glass manufacturing energy, partially offset by the need to make sturdier containers. A further advantage is the substitution of low-temperature heat for washing containers for high-temperature heat in glass manufacture. The former can be provided readily by solar collectors now on the market. Overall saving: 50 percent.

We now summarize our results so far in table 6, which recapitulates the second column of table 5 and applies coefficients of reduced energy use from the preceding text.

Chapter 2 asserts that the 1980 industrial output could have been produced with about 40 percent of the energy actually used. Our examination of industries that used half of all industrial energy has produced estimates which bracket that figure. Is this biased? Does it reflect possibilities not found in the rest of manufacturing, and in agriculture, mining, and construction? Probably not. Industries such as food processing have all the possibilities for more efficient use of heat with the further

Table 6 Energy Reductions for Examined Industries.

Category	Energy use before reductions	Efficiency coefficients	Energy use after reductions
Chemicals	4.0	.4–.5	1.60–2.00
Iron and steel	3.1	.33–.4	1.02–1.24
Aluminum	.9	.5	.45
Petroleum refining	3.7	.26	.96
Paper	2.6	.4–.5	1.04–1.30
Cement	.5	.4	.20
Glass	.5	.5	.25
Total	15.3	.36–.42	5.52–6.40

advantage (see chapter 3) of using heat in temperature ranges well suited to direct solar heat. In the industries that are less energy-intensive in processing, uses like lighting, space heating, and space cooling are relatively more important. Large efficiency gains similar to those found in the commercial sector become available. "Mining" as a category uses much natural gas as a means of extracting petroleum—a use that would decline with petroleum use (depending on how that decline was allocated between imports and domestic extraction). Even feedstocks, a relatively large category, is subject to some reduction. For one brief example: much natural gas is used in the production of inorganic fertilizer. Fully priced natural gas would encourage the production of methane from animal manures and other readily available sources in the biomass category discussed in chapter 3. An inevitable by-product is a nutrient-rich slurry that makes an excellent fertilizer. The relatively decentralized nature of methane generation partially overcomes the transportation disadvantage of bulkier organic fertilizers. For the fastidious, it may be noted that large efficiency gains also are available in the production of artificial fertilizer. We conclude, then, that our sample is not unfair and that it yields an industrial energy reduction of 36 to 42 percent.[81]

3

Renewable Energy in the United States

This chapter examines the potential outputs of the various forms of renewable energy in the United States. These energy sources—direct solar heat, waterpower, windpower, or combustible materials from plant photosynthesis—are continually available. They may be contrasted with energy from fossil fuels, which, once burned, are not again available for human use.

The various renewable energy technologies are surveyed. Each is appraised with respect to its present technical status, its cost, and its stage of commercialization. The technologies are grouped according to their suitabilities to supply electricity, liquid fuels, or heat, since these forms of energy are closely related to the various end uses examined in chapter 2.

The key question is a quantitative one—how much renewable energy can be generated? Only a trifling amount (the "establishment" view) or enough, potentially, to power the economy? The question is sharpened by asking, first, whether the renewable energy sources are capable of providing the 20–25 quads that an energy-efficient United States would have demanded in 1980.

The question "how much" must, for economists, be accompanied by the question "at what cost?" The cost criteria used here are not quite as strict as those used in chapter 2 for conservation investments. There the criterion was the ability of a conservation investment to compete with 1980 long-run marginal supply prices for conventional energy sources, *given existing technologies and costs.* In this chapter renewable energy sources and their potential outputs are assessed according to their prospects, within a decade, of competing with these same conventional energy prices. It is shown that 20–25 quads of renewable energy are available in the United States by these standards.

It is not sufficient to note an energy demand of 20–25 quads and then to cite potential supplies of that magnitude. One must examine the end uses of energy and determine whether it can be supplied in the appropriate forms. That analysis is carried out in this chapter. Many of the high-growth, high-energy scenarios of the 1970s did not get beyond extrapolating out to enormous energy demands, and then listing some equally large number of quads from coal, nuclear power, or whatever source. The inconsistencies of these scenarios were more readily discerned when attempts were made to match sources and end uses.

Renewable energy critics have raised several objections. It is charged that some renewable energy systems produce little or no "net energy"— that more energy goes in than comes out. Since some renewable energy sources are intermittent, energy storage issues must be considered. These two matters—net energy and storage—are considered here.

Last of all, the question of future growth is considered. If 1980's calculated demands could have been met, what about the demands of a growing economy? The conclusion is that fourfold economic growth could be accommodated without encountering sharp rises in the cost of renewable energy.

This chapter is concerned only with establishing the possibility of a renewable energy future—a kind of "existence theorem." The process of getting there—the time path and the requisite policy measures—are treated in chapters 8 and 9.

What, then, is the potential for meeting energy demands of 20–25 quads from renewable resources? There are clearly no technical limits to renewable energy supply at this level or an even greater one. The question is, of course, at what cost? This matter is taken up as we look at each technology in turn. In the section on energy conservation/efficiency we applied the strict criterion of cost-effectiveness against 1980 marginal supply prices of conventional sources. Some of the renewable technologies considered here already meet that test. None are included unless they are known to work already and have prospects of meeting that cost criterion within the next decade.

Renewable energy technologies intercept natural energy flows and use them for human purposes. This has been done for millenia; the only new thing is that modern technologies have higher conversion efficiencies and lower costs. Most of these natural energy flows are derived from solar radiation striking the earth. The resource, though dilute, is certainly abundant—some 44,000 quads fall on the United States each year.[1] Variation between clear, sunny regions and cold, cloudy ones is surprisingly small—annual totals per acre or square foot vary by less than a factor of two.

Solar energy appears indirectly as falling water, blowing wind, or plant matter (biomass). The latter can be burned directly as a fuel or converted into liquid or gaseous fuels. Other, smaller, nonsolar renewable energy flows come from the earth's interior heat (geothermal energy) or from the movement of the tides.

The following discussion of renewable energy technologies, their present status, and their future prospects is intended to be an informational sketch for the reader who is not yet acquainted with these matters. More detailed treatments are readily available.[2]

Useful categories for considering renewable energy supply technologies are those of electricity, heat, and liquid fuels. This might seem like the wrong order, since electricity is essential for less than 10 percent of end-use energy. Heat and liquid fuels are the major components of the "energy problem." Nevertheless, thirty years of research efforts by governments, here and abroad, have been heavily focused on electricity supply. Curiously enough, behind only low-temperature heat, electricity is the energy form most readily supplied from renewable sources.

Renewable Electricity

The renewable electric technologies considered here are: (1) hydroelectricity, (2) geothermal electricity, (3) photovoltaic electricity, (4) other solar-electric technologies, (5) wind-electric systems, (6) tidal power, and (7) industrial cogeneration with biomass fuels.

Hydroelectricity.   Falling water has been used as an energy source for thousands of years and as an electricity source for a century. Hydropower is generally the cheapest source of electricity. Most of the prime large dam sites have already been developed in the United States. Rising real costs of electricity have restored competitiveness to many small sites with dams and have encouraged the installation of generators at many of the nation's dams that were built for other purposes. Refurbishing existing equipment frequently increases electricity output.

Many utilities also have built pumped-storage facilities. During off-peak hours, water is pumped from a lower reservoir to a higher one. Then during times of peak demand, the stored water can be released. This capacity enables utility baseload equipment to run steadily while demands fluctuate. Ordinary hydro facilities also can be used for peak demands. Stream flow raises water in the reservoir during off-peak times, and the turbines operate at peak times.

These facilities, designed to accommodate baseload coal and nuclear generating plants, will turn out to be highly convenient for supply systems with solar-electric and wind-electric components. The reservoirs become the chief storage medium to accommodate fluctuations in output.

Geothermal electricity.   Geothermal steam has been used commercially to generate electricity at Laradello, Italy, since 1904. Low-temperature geothermal resources are being developed rapidly for space heat. Some researchers would not include this resource among renewable energies on grounds that it is subject to depletion. It seems, though, to be capable of producing heat indefinitely if withdrawal rates are carefully established. "Old Faithful" has been releasing heat to the atmosphere for a very long time.

At present, most geothermal electricity comes from dry steam that drives conventional combustion turbines. This is a small fraction of the total geothermal heat resources. A large expansion of geothermal electricity supply therefore depends on technologies that use steam–hot water mixtures, or hot water alone. The resource is even larger if heat from hot, dry rocks can be tapped. Several plants generating electricity from hot water are now running on a pilot basis.

Photovoltaic electricity. Photovoltaic cells depend on properties of semiconductors to convert radiant energy into electricity. Conversion efficiencies have a theoretical maximum of 23 percent; commercially available cells reach 15 percent. This is an important matter, for it governs the space requirements of photovoltaic systems and their ability to meet relatively concentrated energy demands with solar electricity. A square foot in New York City receives about 130 kilowatt hours of thermal energy in the course of a year. It makes a great deal of difference if 15 percent rather than .1 percent can be converted into a usable energy form.[3]

Photovoltaics seemingly have been three to four years away from high-volume commercialization every year since about 1977. They still seem to be some years away from competiveness with conventional electricity sources. They are, of course, already a commercial success for remote installations using relatively little power. The lowest prices reported in 1985 are in the range of $4–5 per watt of peak (midday) power.[4] Most sales are of crystal silicon cells. Several different versions are moving steadily toward commercial-scale production, with probable prices in the $3 to $4 range—one that opens a vast market now filled by diesel generators all over the world. There is much difference of opinion as to the rate at which prices will fall and as to which technology will emerge as the lowest-cost power producer. There is virtually no disagreement with the proposition that PV's will, within the decade, become a competitive source of electricity.[5]

Other Solar-Electric Technologies

There are other ways of converting solar radiation into electricity. The "power tower" in California is an example. A field of mirrors focuses on a boiler atop a tower. A working fluid is heated, which in turn makes steam to drive a turbine. The hot fluid also may be stored to generate electricity at night. The initial plant was very costly, but it seems to work well. Some industry analysts think that it can become competitive.[6] Other solar-electric plants use concentrators (dishes or parabolic troughs) to heat working fluids nearer to the ground than the "power tower." The fluids also make steam to run a turbine. One such plant now operates in the California desert and another is under construction—with considerably lower costs. The latest installation has a capital cost approaching the average capital costs of nuclear plants entering service in 1986.[7]

A simple but promising idea is the solar pond, developed in Israel. If the bottom layer of water is heated by the sun and then prevented from rising (in most cases by the addition of salt, which makes the bottom layer too heavy to rise when heated), then heat can reach usable temperatures and be stored in the pond. Electricity can be generated, or, for that matter, any low-temperature heat demand can be met. Once this principle was discovered, natural solar ponds were found to have been there all along. This is a land-intensive technology with a fairly low conversion efficiency, but it is well suited to arid areas where land has few alternative uses.

Wind turbines.   Wind energy is used to generate electricity, to provide mechanical drive (as in pumping water), or to propel vessels. The latter two technologies have been in use for thousands of years, the former for a century. Most recent efforts have been directed toward improving technologies for wind electricity, though some efforts have been aimed at sail-assisted ocean transport.[8]

Wind electricity from small machines has long been cost-effective in remote locations. Machines in the 30–100 kilowatt class are nearing unsubsidized competitiveness for supplying the grid. Electricity is produced at costs of 12–15¢ per kilowatt hour when the effects of tax benefits are removed.[9] The equipment is still being manufactured in a manner similar to automobiles in the days before Henry Ford introduced mass production assembly lines. Large machines, which have received virtually all of the federal research support, have not yet operated successfully for long periods. Further engineering is required, but there seem to be no insuperable barriers. The California Energy Commission has reviewed the experience with windpower to date and the prospects for cost reduction with further operating experience and mass production of machines. The commission concludes that windpower, along with hydroelectricity, will be one of California's lowest-cost sources of electricity by 1990.[10]

Tidal electricity.   Tidal power is not a huge resource worldwide, but it is locally important in the United States (New England and Alaska), the United Kingdom, France, India, the USSR, China, Korea, Canada, and Argentina. The technology for generating electricity clearly works; a 240-megawatt plant has run in France since 1965. This technology is probably competitive in most places; most North American operating data will soon be available from a pilot installation in Canada.

Biomass-fired cogeneration.   The cogeneration of heat and electricity in industry (and, for that matter, in commercial buildings as well) has been discussed in chapter 2 as an energy conservation technology. It is also a renewable energy source if the fuel is renewable. The most important examples of biomass-fired cogeneration are found in the forest products industry where wood wastes are burned to provide process heat. As new investment projects work their way through the capital budgeting process, electricity generation equipment is being added with new boilers.

Renewable Electricity: Summary

There are indeed many sources of renewable electricity. As they are more widely recognized, and as some of the technologies listed here become increasingly competitive, concern with future electricity supplies will diminish.

As to quantities: hydroelectricity sources already provide 14 to 15 percent of U.S. electricity. More new capacity comes into service each year. Geothermal capacity is approaching 1,500 megawatts, and wind electric capacity exceeds 1,100 megawatts.[11] Both are concentrated in California but are found also in Hawaii and other western states. Wind-power development is also under way in New York and New England. Cumulated production of photovoltaic capacity is small—some 100 megawatts. The significance of this source is its rapid growth and potential for large-scale use in the 1900s.[12]

This summary of renewable electricity technologies underlines the potential abundance of electricity in a renewable energy future. All of the technologies listed are commercialized. All are cost-competitive, except wind-electric (within one to five years), photovoltaic electricity (within ten years), and solar-thermal (probably within five years).

Biomass Energy; Liquid Fuels

Biomass represents another large renewable resource. The term "biomass" has come into use because it is awkward to keep referring to "wood, other plant-based energy sources, and organic wastes."

The principal method now used to extract energy from biomass is simply to burn wood. This may change in the future, since liquid fuels to replace petroleum products will one day be needed. All plant matter can be turned into liquid fuels through various chemical processes. Some plants directly yield, as crops, oils similar to diesel fuel or lubricating oils.[13]

Methane (natural gas) is produced almost any time that decaying biomass is protected from oxygen and the temperature is in the right range. Swamps, landfills, and sewage plants make methane willy-nilly. Modern technology enters to control the process, maximize yields, and reduce costs.

As to biomass energy sources that are presently cost-effective: Scraps and other wood wastes are widely used in the forest products industries for process heat (and, with cogeneration equipment, electricity). Wood use for residential heat also has increased since 1973. Small amounts of methane are now recovered economically, mostly from garbage landfills and sewage plants. The technology exists and works to derive methane from numerous other sources—feedlot manure, dairy manure, food processing wastes, or even energy crops like water hyacinths.[14] Commercialization awaits further increases in natural gas prices and further cost reductions in the processes.

Fuel alcohol (ethanol) made from sugarcane in Brazil and from corn in the United States represents the largest volume of biomass-based liquid fuel now being produced. Production in the United States in 1985 was estimated at 625 million gallons.[15] Large subsidies keep the process viable in the United States, although some new technologies have probably made ethanol viable without subsidies.[16] Most observers, however, think that the most promising large-volume liquid fuel from biomass will be methanol ("wood alcohol"). Research and development activity is aimed at increasing conversion efficiency and lowering costs.

The biomass resource base is quite large in relation to the energy demands of an energy-efficient economy. The most important recent studies are those done by the Office of Technology Assessment[17] and the Energy Research Advisory Board.[18]

Heat

Chapter 2 pointed out the importance of heat in the nation's energy budget. Direct solar heat is particularly well suited to the provision of low-temperature heat—a major use of energy. Space heat in buildings and hot water for buildings and industry can be provided by simple solar collectors. It is these markets that are now being penetrated most rapidly by solar technologies. For space heat in buildings, passive solar design seems more likely to prevail than "active" systems (collectors, pumps, and storage tanks) for reasons of cost, simplicity, and ease of maintenance. By 1983 some 60,000 to 80,000 passive homes had been built.[19] Hot water systems, which necessarily involve collectors and storage tanks, are also headed for simpler designs. Some 1.5 million homes had solar water heat by 1982.[20] Virtually all of the heat used in residential and commercial buildings is in the range that can be provided by solar sources. The major exception is cooking, and that is a relatively small use compared to space heat and water heat.

Higher-temperature solar collectors now on the market can provide heat for industrial applications in the 120°–550° Fahrenheit temperature range. Collectors use parabolic dishes or troughs to collect solar radiation.

Solar heat for low- and medium-temperature industrial processes got a bad name from government-subsidized installations in the 1970s. Since there is no inherent reason why solar collectors cannot work, we are dealing again, not with insuperable barriers, but with familiar problems of design and cost. Aided by tax credits and the same generous depreciation allowances now widely available to business, several commercial, privately financed installations are now under way in such industries as textiles and ceramics. The present installations probably will disappoint investors even if there are no technological problems. Heat sales prices are tied to fossil fuels prices, which are not likely to escalate as rapidly as prospectuses assumed. Nevertheless, it is important to go through the experience of finding workable and durable designs and materials. The present generation of equipment may have reached this point; we should know in a year or so. If more work is required, that is what engineers are good at.

The energy "establishment" seems to have double standards here. On the one hand, it exhibits near-infinite patience with breeders and fusion. It tolerates, even advocates, billions for research, development, and demonstration, through whatever number of iterations may be required. Yet one round of poorly designed industrial collectors fails, and we are assured that such applications of solar energy cannot produce much energy, at least until sometime in the distant future. Prospects for solar industrial heat are appraised in the *Annual Review of Energy, 1984*.[21] The author is basically optimistic. He points out, however, that two- to fourfold cost reductions in industrial solar systems will be required to make this source fully competitive.

That observation may be qualified in two ways. His criterion of competitiveness (quite properly, for short- and medium-term purposes), was cost per million btu's provided, compared with heat from oil, coal, and natural gas. If one were to make the comparison with heat from electricity, many such systems would already be competitive. Heat from fossil fuels ranges in cost from $3.00 to $6.00 per million btu's, with increases to $10 to $12 in prospect over the next decade or so.

Electricity at current rates would provide industrial heat at $15 to $18 per million btu's, and at long-run marginal supply costs, $30. This is an important point, since high-energy, high-electric, high-nuclear futures count on a substantial penetration of industrial heat markets for electricity.

A second qualification has to do with interest rates and solar heat costs. Renewable technologies have high capital costs and, in most cases, free "fuel." The return of real interest rates to levels historically experienced would lower solar heat costs. This development would, of course, also lower electricity costs, but by one-half to two-thirds of the drop in solar heat costs. Another positive appraisal comes from the director of the Solar Energy Research Institute under the Reagan administration—an administration rarely cited for its excess enthusiasm for solar energy. The director, H. M. Hubbard, sees industrial solar process heat moving into commercial maturity without subsidies in the period 1980 to 2000.[22]

This listing, I believe, captures all of the renewable energy technologies that are now commercialized or are likely to be in the next decade. In fifty years this list may well appear to be quaint and obsolete; one never knows what attractive possibilities await discovery.[23] We stick with the known in this study; if something better comes along, that merely strengthens our case.

Present Renewable Energy Output

Renewable sources already provide 6.8–7.5 quads, which is 9 to 10 percent of the U.S. energy supply. Some 3.8 quads come from hydroelectricity and 2.6–3.5 from biomass. This energy is mainly from wood wastes used for process heat and electricity in the forest products industries and from wood burned in households. Only a small fraction of biomass energy is regularly reported in official statistics—that used by utilities to generate electricity. Small amounts of direct solar (mainly water) heating, geothermal electricity, and wind electricity probably add up to about .2 quads of primary energy equivalents.[24]

Table 7 U.S. Renewable Energy Supply Possibilities.

Source	Quads
Hydroelectricity	5
Geothermal electricity	1
Wind electricity	1
Photovoltaic electricity	1
Biomass	9
Direct solar heat	7
Other	1
Total	25

Source: Author's estimates.

Renewable Energy for a 20–25 Quad Economy

This book began with several questions. The second of those questions asked whether renewable energy sources could supply an economy like that of the United States in 1980 if it first had been made energy-efficient. The answer is clearly "yes." Table 7 shows one possible combination of renewable energy sources that would provide 25 quads of primary energy. The quantities for each renewable technology are selected on the basis of very rough estimates of costs at the margin. The lower U.S. figure of 20 quads would be reached even more readily.

"Other" energy sources would probably include .5–.8 quad of metallurgical coal, which not only provides heat but enters chemically into the reduction of metallic ores. An energy-efficient steel industry in 1980 would have required this input, the amount from the range given depending on how much additional recycling of scrap might be achieved. Heat and electricity from solar ponds and tidal electricity also would be included under "other." All of these renewable sources except hydroelectricity are far below their technical limits.[25] This allows much flexibility in meeting the various end-use energy needs.

Of the technologies that are not already fully cost-competitive with conventional fuels at marginal supply prices, only six technical developments are required to make them so: (1) realization of already-projected cost reductions in photovoltaic systems; (2) enough experience with wind turbines to settle on workable and durable designs so that mass production can get under way. (This may now [1986] apply only to large wind machines; it may already have happened for small and medium-sized ones.); (3) cheaper processes for liquid fuels from biomass, especially for methanol;[26] (4) more experience with the design and operation of solar devices for medium- and high-temperature industrial heat; (5) more experience with the design and operation of geothermal electric conversion from hot water and water-steam mixtures; and (6) further process research and cost reduction in biomass-to-methane technologies.

The SERI study mentioned earlier projected 12 to 22 quads from renewable sources by the year 2000, as do many other serious investigators.[27] Given still more time, estimates of much higher amounts—in the 50–70 quad range—are not uncommon.[28] I cite these merely to show that my figures are indeed well below the nation's potential, at prices that are already competitive with marginal supply prices of conventional fuels, or give strong promise of so being within the next decade.

We now have shown that overall renewable energy supplies are adequate in amount to have produced the 1980 GNP. Several refinements are needed. It is important to know whether renewable sources can be matched properly to end uses. It is also important to assess the net energy output of renewable systems. Some critics contend that net energy outputs are small or perhaps negative, when energy embodied in the equipment and operating inputs are considered. When some sources are intermittent, storage of energy is also an important issue.

Matching Sources to End Uses

Electricity would be required for about 8 quads[29] (primary energy equivalent, as the United States uniformly measures it) in the 25-quad total— lights, electronics, electrolysis, and mechanical drive. Table 7 directly identifies 8 quads of electricity, of which hydroelectricity is the largest component. Transportation would require about 6 quads of liquid fuels, less any reductions from electrified rail transport and shifts from truck freight to rails. The rest is heat at various temperatures. Renewable electricity and low-temperature heat are relatively abundant; liquid fuels for transportation and industrial process heat will be our chief concerns.

Heat sources.   Space heat, a low-temperature use of heat, now consumes 17 percent of the nation's energy budget, but that energy demand will virtually disappear in an efficient economy. Hot water in residences and commercial buildings (now 3 percent of national energy use) would be reduced as a claim on energy inputs through conservation measures. Most of the remaining demand would be met with direct solar heat, or perhaps by geothermal heat. This leaves medium- and high-temperature heat demands, mostly in industry, but partly for such residential and commercial uses as cooking and clothes drying.

After all of the efficiency measures are taken, there would remain some 4 to 5 quads of industrial heat demand and perhaps .6–.8 quads in the other sectors, or 4.6–5.8 quads altogether. The use of coal in the steel industry reduces this to 4–5.3 quads.

Much of this can be supplied by direct solar heat or, as it is now, by the direct combustion of wood wastes.[30] For the solar portion, some back-up sources are required. Enlarging the back-up share would reduce the requirement for heat storage and thus hasten the time of cost-competitiveness.

Some of the heat energy can be provided by extending the use of wood or wood wastes to industries outside the forest products group. In fact, this process is already under way in wood-rich areas. Much of the biomass resource (about 2–3 quads) is in forms that are readily converted to methane. Thus, about 1–1.5 quads (allowing for conversion losses) would be available for very high-temperature heat demands.

This volume of gas fuel would not suffice to sustain a gas pipeline system constructed to handle some 20 quads (the present size). We would be more likely to see smaller regional pipelines linking methane producers and industrial users, or a clustering of industrial users near biogas-methane sources.[31]

Methanol and other liquid fuels also would be available as an industrial fuel. There would appear, then, to be ample renewable resources to provide heat in the quantities indicated above. We now have direct solar heat with storage and direct combustion of wood or other biomass, methane, or biomass-derived liquid fuels. We do not know enough about the relative prices that may emerge as conversion technologies mature to say any more about the roles of each form of energy. It suffices to show that some combination of suitable renewable sources would be adequate.

Transportation and liquid fuels.   In the transportation sector liquid fuel is now the overwhelming choice. We see from chapter 2 that the estimated transportation energy demand in an energy-efficient economy would be 5–6 quads. A small portion of this is already provided by electricity—some used in urban subway or rail systems and some on the railroad network. This use would be increased by some further electrification of the rail system and at the same time reduced by rail efficiency gains.

The development of cheap, light, long-lived batteries through many charge-discharge cycles would greatly alter the transportation fuel situation. That development, so essential for high-electricity, high-nuclear energy futures, is not considered here, though it would be a most welcome development. (Auto battery-charging would be ideally matched with the use of intermittent electricity sources—photovoltaic cells or wind generators, for example. This would extend their potential contribution to the total energy supply without further independent storage.) In the absence of such batteries, about 5 quads of liquid fuels must be provided for autos, trucks, airplanes, and other carriers. The renewable source is, of course, biomass.

Much attention has been focused on ethanol (grain alcohol), which is presently being produced at a rate of about 625 million gallons per year. This program is heavily subsidized through the remission of federal gasoline taxes and many state taxes as well on blends containing ethanol. This effort has been criticized as putting fuel crops and food crops in direct competition and yielding little net energy after energy inputs into corn, distilling, etc., are considered.

The net energy aspect of the matter is discussed in the next section. Present practices, after all, developed over the centuries for the production of alcoholic beverages, not fuels. Technology is not helpless in these matters. On the food-fuel competition issue, it is estimated that ethanol production in the United States can reach three to four billion gallons before there is any appreciable problem.[32] Corn ethanol yields a by-product, distiller's dried grain, which is a protein-rich cattle food supplement. Soybean demand is thereby reduced, partly offsetting the acreage devoted to additional corn. The animal feed market for distiller's dried grain saturates at the production level of three to four billion gallons.

A portion of the liquid fuel supply, then, perhaps .2–.3 quads, could be supplied from ethanol feedstocks without undue competition in food acreage. Processes that derive ethanol from wood and other ligno-cellulosic plant matter are being developed. The ethanol resource base could become much larger.[33]

Plants that directly produce petroleumlike liquids also are being investigated.[34] Some of them grow well in arid or semiarid areas, so that they would not be in direct competition with food or forest production. Many of the investigators involved in these projects believe that large-scale supply may be feasible. It is too early to pass judgment on these views. If they are right, liquid fuel problems are the more easily solved. Even if they are wrong, some fraction of fuels, lubricating oils, and industrial feedstocks can come from this source.

Fortunately, there is another quite large liquid fuel source—methanol. Most investigators who are looking at liquid fuel sources other than coal, tar sands, or shale seem to believe that methanol ("wood alcohol") is the most promising candidate, at least in the light of presently available information. It is already produced in commercial quantities from natural gas. Indeed, the first new methanol sources are likely to be sites where gas is now flared. Developing nations with gas sources remote from industrial centers may find it cheaper to produce and transport methanol than to liquefy the natural gas. The latter process involves high-pressure, very cold processes and shipping equipment, with equally complicated and dangerous receiving and regasifying facilities. Transitions to renewable fuels are always easier when there is an established product and where the renewable fuel fits into an existing system. Renewable sources can then gradually displace the nonrenewable ones.

Methanol can be made from wood, or from virtually any form of plant life—food-processing waste, the organic components of garbage, or yard trash. For these reasons methanol now looks like the most promising source.

Methanol is presently cheaper than gasoline, but, as a transportation fuel, it faces some formidable market barriers. Auto and truck engines now in use will not switch back and forth between gasoline and methanol. Engines must be modified to run on methanol. People will not buy methanol-powered vehicles until there is a fuel supply system, and suppliers will not create a system until there is an ample market—some critical number of vehicles using the fuel.

California is now attempting to break out of this impasse by simultaneously encouraging methanol pumps at service stations along with state and private fleets of vehicles that use it. The federal Department of Energy is also working toward the creation of methanol-powered vehicle fleets and a compatible fuel distribution system.

My analysis of liquid fuels, then, is conducted primarily in terms of methanol. This fuel, perhaps supplemented by ethanol and oil from plants, would have to fuel the transportation sector. Once the efficiency gains discussed in chapter 2 are achieved, the liquid fuel demand remaining is about 5 quads. Since this area is thought to be one of the most severe challenges for a "solar" economy, we shall examine the fuel supply in some detail.

Methanol, per gallon, has but half the energy content of gasoline. Still, it yields more than half as many miles per gallon since it burns more efficiently.[35] This additional mileage is 20–25 percent.[36] Therefore 5 quads of petroleum-based fuels are replaced with 4–4.2 quads of methanol. The amount of wood or other methanol feedstock required depends on the efficiency of the conversion process. Typical estimates of this conversion factor in the energy industries are in the range of 35 to 40 percent.[37] Those seriously pursuing higher efficiencies seem to find them; one process reportedly reached an 83 percent conversion efficiency.[38] The SERI study used 70 percent as an estimate of an attainable commercial-scale process; another authority uses 60 percent.[39]

At a 70 percent figure (70 percent of the energy content of the feedstock emerging as energy content in the methanol), then, 6 quads of wood, wood waste, cotton gin trash, or whatever might be at hand would be required. Since methanol plants can be built on a small scale to serve localized feedstock sources, transportation allowances would be modest as an energy charge against the process.

This analysis suggests a small upward revision in our primary energy requirements, since conversion losses must be considered. When crude oil is processed into gasoline, about 10 percent of the energy content of the crude is used as a processing fuel. Since refining is considered to be an industrial activity, this energy use is part of the industrial sector total. In 1980 about 1.5 quads of liquid fuel were used in petroleum refining attributable to transportation. When we reduced oil use by two-thirds with respect to increased efficiency, we took 1.0 quad out of the energy use total for industry. If 6 quads of biomass feedstocks (energy content) are required to replace 5 quads of gasoline and other petroleum fuels, then primary energy inputs will have to allow for this. The addition is .5 quad (1 quad in producing methanol, less the .5 quads already allowed for in the refining of petroleum). Part of this, incidentally, reflects the present importation of refined products; the energy used in refining is charged to some other nation.

The real issue here is the adequacy of the biomass resource base for transportation fuels, bearing in mind that some biomass will be used in industry and, to a much lesser degree, in residences. If the gross potential is 17 quads (18 counting municipal solid waste), then, at this level of use, there would appear to be an adequate biomass resource. Many biomass resource estimates, incidentally, omit much material remote from markets. The use of small-scale methanol producing units, perhaps even portable ones on barges, rail cars, or trucks, would enlarge the usable fraction of remote biomass sources.

This discussion has shown that the end-use demands for various forms of energy in a 20–25 quad economy can be matched with renewable sources. That would be very difficult at 75 or 100 quads; again, it is the demand reduction through conservation-efficiency investments that is the key element.

"Net Energy" Critiques

One issue that renewable energy advocates must face is that of net energy output. Many critics have argued that some devices produce little, if any, more energy than is consumed in their manufacture and installation. One such critic is Petr Beckmann;[40] we shall use an example he cites as a point of departure.

He cites a study[41] alleging that solar collectors in an average U.S. climate return only about 2.8 times the energy used in making and operating them. Energy-intensive materials like aluminum, copper, and glass (or clear, hard plastic) are typically used. Then there is fabrication, transportation, and installation, all of which add some energy. Electric pumps also use energy for operations.

We then correct the estimates in the following ways:

1. When energy-efficient processes are used in the materials production industries, "embodied" energy in the materials drops about 30 percent (see chapter 2, appendix).

2. Copper and aluminum may be recycled when the useful life of the collector is ended. The recycling "credit" for aluminum is about 95 percent (energy use in recycling vs. virgin aluminum) less the inevitable materials losses. Averaging across the various materials, about half the materials employed in energy production can be "reused."

These two adjustments reduce the use of "embodied energy" by 65 percent.

3. The example given tried to heat, with an enormous and complex system (panels, pumps, and storage tanks), a poorly insulated house. If one concedes that space heating is better done with superinsulation or passive solar design, then we are down to water heaters. Storage may not be charged to the solar system except as it exceeds in size the storage that would be there anyway. "Parasitic" losses (to operate pumps) are more like 7–10 percent of the energy produced, not the 20 percent used in his example.

When all of these adjustments are made, energy outputs from solar panels exceed inputs by about eight to ten times—a ratio not far from those given for fossil and nuclear power plants.

The preceding discussion dealt with the "net energy" issue as it pertains to solar collectors. Another point at which the question has been raised is that of grain alcohol. It takes energy to cultivate, fertilize, and harvest corn, and still more to convert it to pure alcohol. Some early operations did indeed seem to use more energy, even oil or gas fuel, than that which emerged in the ethanol. More efficient processes seem to have allayed this concern.[42]

The criticisms of renewable energy devices on net energy grounds, and the responses to them, are usually framed in technical terms. Indeed, we have just shown that two critiques of renewable technologies can be effectively rebutted.

One of the useful insights of economics is that there is a much better way in which to frame the questions and the responses. Useful energy comes in various forms. Some forms command higher prices than others. Systems that convert low-value energy into high-value forms may lose much or little energy in the conversion process; that is beside the point. Systems will not survive in competitive markets and without subsidies unless the value of the outputs exceeds the value of the inputs, including the energy inputs. Coal-fired generating plants lose about two-thirds of the energy content of the fuels, but they nonetheless may be viable if the price of electricity is sufficiently high. Energy systems with net losses not covered by sufficiently high prices for energy outputs will not survive.

In this light, it does not matter whether solar collectors have lower or higher net energy outputs than nuclear power plants. They are economically viable if they are the lowest-cost source of low-temperature heat.

Many of the critics are making an implicitly economic argument if they assert that solar heat or ethanol would not survive without subsidies. That must be determined on a case-by-case basis. Their arguments would be more evenhanded if they were directed with equal vigor to nuclear power, which is also highly subsidized.

Energy Storage

Another issue that must be considered in renewable energy supply scenarios is that of energy storage. The Union of Concerned Scientists study mentioned in chapter 1 gave a great deal of attention to this matter. It was somewhat more important there, since that was a higher-energy scenario than mine. The problem is diminished at lower energy demands, but nonetheless it is real.

Biomass fuels, over one-third of the energy input listed in table 7, already represent a kind of stored energy, as is the case with fossil fuels now used. The storage problem, then, is one for electricity and heat.

Wind and solar electricity are thought to be intermittent (which they are) and therefore unusable in volume without enormous storage capacity—requiring, say, a technological breakthrough in battery or other storage. Up to about 2–3 quads each, this is not so. Integration into grids with 5 quads of hydroelectricity, further pumped storage capacity, some geothermal baseload capacity, and, for good measure, biomass cogeneration and solar pond contributions converts electricity storage into a nonproblem.[43] Electricity storage becomes critical only at higher levels of usage, levels not likely to be demanded. Wind generators, when grid-connected, are capable of meeting some baseload need (i.e., do not require full conventional backup). Model-makers have shown that, given sufficiently dispersed sites with at least partially uncorrelated wind regimes, a wind grid alone can approach a conventionally powered grid in reliability.[44] Photovoltaic electricity is predictably intermittent with respect to day-night and unpredictably intermittent with respect to cloud cover. Given the partial inverse correlation between sunlight and wind speed in many areas, and given the storage capacity of hydroelectric[45] reservoirs, a renewable-based, multisource grid has, at worst, a problem of shifting loads into peak sunlight hours. On the demand side, air-conditioning loads are partially correlated with peak photovoltaic output. Storage of "cool" in buildings can bridge the gap between peak power production at midday and peak air-conditioning loads that occur later in the day. The seasonal correlations are too obvious for comment.

On the issue of electrical storage it also should be pointed out that high-energy, high-electric, high-nuclear scenarios are the ones with truly massive storage problems. Nuclear plants are baseload plants. A million or so kilowatts of generating capacity is either on or off. Loads vary enormously, daily and seasonally. Load smoothing helps in all scenarios but gets more difficult as more end-use activities are electrified. Extending electricity markets requires, for example, space heating—the very activity that exacerbates winter peaks for utilities. Developments in storing electricity, which are absolutely essential to most official energy futures (in that they tend to be high electric and high nuclear), are helpful to renewable-energy futures as well. The reader should note the difference between "helpful" and "essential."

Another interesting parallel pertains to electric vehicles. They (and their attendant storage batteries) are essential to nuclear futures, since electricity must substitute for liquid fuels. Off-peak battery charging is highly desirable for load-smoothing. These two characteristics of electric vehicles mesh neatly with the needs of renewable energy futures; biomass-based liquid fuels could be limiting factors in some countries, depending on vehicle efficiencies achieved and biomass availability. Vehicle batteries could be charged from intermittent sources like photovoltaic cells, thus expanding their contribution.

Storage of heat is the other category for the discussion of energy storage. For most low-temperature needs, storage is already available in existing systems. Residential and commercial hot water, for example, already use storage tanks. Larger tanks may be warranted to extend storage periods. Insulation of storage tanks beyond 1980 standards is likely to be cost-effective as a conservation measure anyway. Comparing enlarged storage with backup systems is a straightforward economic calculation. Using long-run costs for fuels and electricity uniformly raises the solar fraction of total energy provided and enlarges storage. When cooling is called for, "cool" may be stored in building mass, in ice, or in other media.

Storage at higher temperatures also is possible. The principles are the same—insulated tanks—but the medium may be different—oils or other working fluids, rather than water.

Very high temperatures such as those used in smelting ores do not lend themselves readily to solar heat. Nevertheless, there are many opportunities for solar preheating. The biomass resource is sufficiently abundant to cover this heat demand in an energy-efficient economy. In cases where carbon enters directly into the chemical reaction, as in reducing iron ore, continued use of coal has already been suggested.

When the energy storage question is raised for a low-energy economy, and when it is broken into its component parts, the difficulties are much smaller than might be supposed. Biomass fuels (one-third or more of the renewable energy economy) store energy as readily as fossil fuels. Electricity, another third, has a vast storage system already in the nation's hydroelectric and pumped-storage facilities. Low-temperature heat is readily stored. High-temperature industrial applications are more likely to be met with biomass-based fuels than with direct solar process heat. Storage components of solar heating systems for intermediate-range industrial process heat—perhaps 10 to 15 percent of the nation's energy—is the area in which some additional research and development may be needed.

Growth

It should now be established, subject to the reader's checking the references, and using his own judgment where judgment, not facts, have been used, that the U.S. 1980 GNP could have been produced with an energy input of 20–25 quads, which in turn could be supplied from renewable resources. We turn now to the next issue discussed in the introduction, that of the adequacy of the renewable energy inputs in an economy growing over time.

This is done in part to make a clear analytical separation between issues of energy efficiency and renewable energy on the one hand and economic growth on the other. To be sure, many solar advocates have reservations about continued economic growth. The two sets of views are not necessarily indivisible. One may favor an energy-efficient, energy-conserving future and also the use of renewable energy (these also are, of course, separable issues), while favoring, opposing, or holding no opinion about future economic growth. It is only the discovery during the last decade of the enormous potential for energy efficiency gains that makes this separation of issues possible.[46] If energy-GNP coefficients were fixed or nearly so at 1973 levels, a low-energy future would necessarily be a low-output future.

My approach is different from that of the other studies cited. They all begin with the year of the study and then combine, over a period of years, a gradual "mining" of potential efficiency gains (which, cet. par., reduce energy demands) with growth (which increases them). SERI, for example, lets GNP grow by 80 percent from 1977 to 2000, while energy use drops from 76 quads to 62–66. Efficiency gains more than offset growth.

Here we are analytically separating the two processes. We first estimate *all* of the efficiency gains available with existing technologies that are cost-effective. Then we ask how this much-reduced energy demand might grow over time.

Any energy future (indeed any future) that we would want for our progeny must entail a slowing down and eventual ending of population growth. This slowing, fortunately, is already taking place all over the world. The failure to notice it earlier accounts, in part, for so many enormous overestimations of future energy demands in the developed nations. Population has already reached a plateau in much of Europe. It will probably do so in the United States sometime in the next century; it would do so sooner had Latin America awakened a generation earlier to its population crisis, for it is immigration, mostly illegal, that now confounds U.S. population estimates. A stable population does not imply the end of economic growth. Output per capita might rise indefinitely (I rather think not, but my case does not rest on that eventuality).

The central issue then becomes: what is the nature of the relationship between economic activity and energy demand when the latter has been pushed down to the levels already described? Recent experience or SERI-type projections are of little help. Recall that in the United States from 1973 to 1979 real GNP grew at an average rate of 2.7 percent while energy use grew at .9 percent, suggesting a marginal growth coefficient of .33. In Japan, with higher growth rates both in GNP and in energy efficiency, the coefficient was similar. But these data are not relevant to our question. In both cases some of the existing potential efficiency gains were being realized while output was rising. We learn nothing of the relationships that might have obtained had either economy started from a highly efficient point.

I think it can be argued credibly that the marginal energy-GNP coefficient is less than 1.0 in developed, energy-efficient societies with stable populations. I am willing to make the even bolder assertion that the coefficient will approach a .0 asymptote in a generation or so.

The reasoning behind this assertion is contained in the following three propositions:

1. *Saturation.* All goods and services are subject to this saturation. Growth continues only because new goods and services emerge. Empirically, the most energy-intensive goods and services are at or nearing saturation (the sole exception that comes to mind may be air travel). Only one vehicle can be driven at a time by each person of driving age. Even in a very wealthy society there would be some adults who would not choose to own and drive vehicles (the imprisoned, the disabled, some of those who live in Manhattan, for example). Not all drivers will spend rising numbers of their waking hours driving as their incomes rise.

As to space heating, people do not seem to wish to be hotter and hotter as they grow wealthier and wealthier (though nearly every high-energy demand forecast implicitly assumes this—and nuclear-electric space heat, at that). The same observation applies to space cooling. Space conditioning energy demands eventually would rise only with square footage of space occupied, which does rise with income to some degree, but not indefinitely.

These examples deal with energy use in transportation and in buildings. What of the industrial sector? The materials component of U.S. GNP has been declining for some time. Materials are the most energy-intensive component of industry; one would expect, therefore, that industrial energy use would rise in the future less rapidly than GNP. Moreover, some processes will no doubt move abroad (or, more accurately, trends in that direction are likely to continue). The clearest example is perhaps that of alumina; the cost of electricity is the decisive factor in location. Further moves to Canada and Brazil, where vast amounts of relatively cheap hydroelectricity are found, will no doubt take place. Within the United States, Alaska will probably become the prime location for alumina production. This might increase U.S. energy use above my lowest estimate, but it also would bring more hydro capacity into use.

2. *Technology.* There is no reason to suppose that technological advance in energy efficiency will cease. Indeed, there is reason to expect that the pace will quicken as more scientific and engineering talent is brought into this activity. Technical change has been successful in raising labor productivity for two centuries, thus enabling real wages to rise. There is no reason to expect less for the energy inputs in a period of secularly rising energy prices. Growth models and a vast empirical liter-

ature on growth are instructive: without technical change and with a constant labor force, growth soon ceases anyway. No violence is done to the growth assumption by extending some of the requisite technical change into energy input areas.

In considering technological change we must also consider the emergence of new, energy-intensive goods and services. If we had been, in 1890, considering future energy uses, amazing prophetic powers would have been required to contemplate 1980's energy use for autos, airplanes, and air-conditioners. About such unknown and unknowable developments, one can say little. We here assume that technology is many-faceted; new goods and services, some energy-intensive, will probably emerge. So also probably will new efficiency gains and new renewable energy sources. No one can say that any one of these developments is any more likely than another.

3. *Relative prices*. To the extent that saturation and technology do not suffice, some combination of rising energy prices, price elasticities, and income elasticities will hold energy demands in a range that can be met from renewable resources. Suppose, for example, that real GNP grew at 3 percent per year, energy prices rose at 3 percent per year, long-run price elasticities were -1 and income elasticities were $+1$. Zero energy growth would hold with *no* technical change. The role of technological advance in energy efficiency or energy supply is thus seen to be that of moderating energy price increases.

Now the nagging question: couldn't the whole process someday be crippled by scarce energy inputs? Couldn't there be some discontinuities in production functions, or some supply limits suddenly breached, un-awares, to that growth is halted while the least well-off are still in poverty? Some such tacit assumptions seem to underlie the sense of urgency displayed by many energy analysts.

Perhaps, but not for the next century or so. As evidence we first cite the very high price and substitution elasticities implied by the estimates in chapter 2. On the renewable energy supply side, I shall next try to demonstrate that costs do not rise markedly in the 25–50 quad range (though they might well beyond that range). As a crude example, if the energy-GNP coefficient at the margin remained at .5, national output could quadruple with a doubling of renewable energy inputs from 25 to 50 quads.

Should energy demands rise beyond the 25-quad range, additional

Table 8 Expanded U.S. Renewable Energy Supplies.

Source	Quads	Comments
Hydroelectricity	6	(Assumes development of some of Alaskan potential, small hydro sites in the lower 48 states, upgrading of existing stations, and continued imports from Canada.)
Geothermal	4	(CONAES "enhanced supply" scenario[47] for 2010.)
Wind	6	(Some additional pumped storage is required
Photovoltaic	6	unless cheap battery storage or fuel-cell type PV cells are available.)
Biomass	18	(Full sustainable potential per Office of Technology Assessment, plus 1 quad from municipal solid waste, not included in OTA estimate.)[48]
Direct solar	9	
Other	1	(Includes metallurgical coal.)
Total	50	

renewable resources come into use according to relative prices, costs, and the nature of end-use demands—electricity, heat at various temperatures, and liquid fuels. Table 8 gives an estimate of renewable energy outputs summing to 50 quads.

This is still less than the 70–80 quads in the "technical limit" scenario of the CONAES Solar Resource Group (actually, their 60 quad plus geothermal and hydroelectric estimates from other CONAES panels). At these levels, limits are being strained. Matching sources to end uses in the right places becomes more difficult. Further increases, especially for biomass-based fuels, would be subject to sharply rising costs, and storage costs would begin to constrain electricity supplies.

These observations explain, perhaps, why energy policymakers have had so much trouble taking renewable energy seriously. They have imagined futures when more energy than ever will be used, and they then have observed (correctly) that such energy output levels would be difficult or impossible to meet from renewable sources. That has turned out to be the wrong question. One must first ask what gains are possible in the efficiency with which energy is used. The answers, developed in the past few years, contain surprisingly good news. Only then does a long-run renewable energy future begin to look feasible.

Suppose, then, that the following propositions have been demonstrated:

1. Energy use in the United States can be decreased greatly, and at costs below those required by new energy supplies.

2. Renewable energy sources have the potential to meet these demands at competitive prices.

3. Economic growth is not thereby energy-constrained for the foreseeable future.

Indeed, that is exactly what has just been demonstrated (pending only six developments in renewable energy supplies), unless there are errors of fact, of reasoning, or of interpretation in the preceding analysis.

This demonstration, which is not widely accepted, is but the first step. It is an essential step; no one can be expected to assist in the transition to a state that he believes to be impossible or nonexistent.

The second step has to do with the transition process itself. That process, and the policies appropriate to it, are taken up in chapters 8 and 9.

4

State, Regional, and Local Issues

We have shown that the United States, taken as a unit, has enough renewable energy to have produced the 1980 output and to provide for economic growth for some time. But what does that imply for states and regions? They are not all equally well endowed with renewable energy sources, just as they are not equally well endowed with present energy sources.

We also must deal with another issue in the use of renewable energy—that of providing energy to areas with dense populations and concentrated economic activity. Some critics of renewable or solar energy, citing its dilute nature, have flatly declared that it cannot deal with such areas. Some other (i.e., nuclear) source, they say, must be found.[1]

Why should it matter? After all, much transportation of electricity and mineral fuels now takes place, and over very long distances. Cannot the same be done in the kind of economy we envisage?

That is not clear at the outset. Fossil fuels are readily transported. They have high energy contents relative to their weights. Electricity, the premium energy source, is also easily transported. Heat, on the other hand, is best used close to its source and can only be transported a few miles in the best of circumstances. In a common view, solar energy = solar collectors = heat, which must be used at the same site or very close by. Another renewable energy source, wood, is heavier than coal or oil per unit of energy. Transport, while technically possible, is costly. A national showing of the technical and economic feasibility of an efficient, renewable energy future is not necessarily complete. We have the additional task of showing that local and regional self-sufficiency obtains, or that regional surpluses can be matched up with regional deficits in the forms of cheaply transportable "energy carriers."

Some location decisions depend on relative energy prices among different regions. Changes in prices of heat, electricity, or liquid fuels may well change location patterns as we compare two states of the economy—the 1980 GNP produced as it was, and the same output as we have shown that it might be produced. Aluminum reduction, for example, is located near cheap electricity, typically hydroelectricity. The location of steel plants is governed by transport routes for ore and coal.

Texas, Louisiana, California, and several coal-producing states are now net energy exporters; most other states are net importers. The changes we are proposing would make all areas more nearly self-sufficient. If any energy is locally produced, reducing energy use by a factor of three or four would raise the local share. As we shall see in more detail in a moment, some renewable energy is located at any point where people are; some local production wlll necessarily take place. Exportable surpluses will probably be in the forms of electricity and liquid fuels from biomass sources. States like those in the Southeast that now import fuels may begin to export instead, since they have large forest resources per capita. These changes would surely affect the locations of many activities besides the energy industries themselves. We note those effects here but do not try to trace them farther. These are, of course, big issues for the parties concerned.

We should make another point that is related to regional and locational matters. It is not really quite accurate to talk about producing the "same" GNP with energy conservation and with renewable energy. The energy sector itself will be much smaller—by a factor of three or four. We should perhaps speak of the "nonenergy sector GNP" as the collection of goods and services we are trying to get produced. Even this is not accurate. In place of oil, gas, coal, and nuclear electricity, we would be using wind turbines, photovoltaic cells, solar collectors and concentrators, and various equipment to convert biomass into usable fuels. Only hydroelectricity, geothermal electricity, and wood-fired boilers from the present energy system would be involved. The industries that supply the energy sector, then, would have quite different outputs. Any attempt to reconstruct a hypothetical GNP, perhaps using an input-output table, would imply more precision than exists. We shall, therefore, note this set of effects, but without further exploration. We also shall continue to suppose that discussing the "same" national output under two quite different energy arrangements, while not accurate, is a workable approximation.

This chapter, then, deals with the issues just listed. First, we examine each region of the United States with respect to its potential production of renewable energy: electricity, liquid fuels, and heat. Then we examine conservation or renewable energy studies that have been developed for California, the Pacific Northwest, New England, and New York City.

The New York City case is used to examine the claim that solar energy is too dilute to support concentrations of population and economic activity. That claim does not appear to be valid. Conservation can reduce energy demands in New York City to about one-third the present level. The reduced demand can be supplied in part from local renewable energy sources—20 to 25 percent. The rest, in the form of electricity and liquid fuels, can be imported from surrounding areas. The volume of imports, incidentally, would be about one-fourth the present level.

Regional Renewable Sources: Electricity

With respect to electricity, the greatest diversity of sources is found in California and in the Pacific Northwest. Geothermal resources are of sufficiently high grade to generate electricity; wind-electric potential is plentiful; and hydroelectricity is abundant. Hawaii is similarly well endowed. These areas also have potential sources in photovoltaic cells and in biomass-fired cogeneration, as do all other areas. Alaska is so abundantly endowed with hydroelectricity that wind and geothermal sources are almost redundant.

Moving across the continent, New England and New York have considerable hydroelectric resources, wind-electric potential, regionally unique access to tidal power, and proximity to Canada's vast hydroelectric resources. The chain of states lying along both sides of the Appalachian ranges have both hydroelectric resources and sites for pumped storage. Some wind potential is found in the mountains. Forest biomass resources are relatively large per capita in most of these states. Florida, which lacks wind, hydroelectric, and geothermal resources, has biomass production rates per acre well above the national average since the growing season is long. This suggests more biomass-fired cogeneration than might be found, say, in New England. Florida is also the only state in the continental United States to have access to baseload electricity from OTEC (Ocean Thermal Energy Conversion). We have not included that particular technology in previous listings, since it is not as mature as the others. It takes advantage of temperature differentials between warm surface waters and deep cold water in the ocean. Those temperature differentials are found quite close to the Florida coast. Hawaii is the only other state where this resource is potentially available.

The Midwest has a large wind-electric potential, especially on the Great Plains. Hydroelectric resources are also abundant,[2] though the areas with concentrated populations may also need to draw on imported Canadian hydroelectricity. The Texas coast is another area with wind-electric potential. The shrinking hydroelectric potential as we move westward is offset by the possibilities for solar ponds. They are space-intensive, not a negative factor in desert areas. Geothermal resources also extend into this area, as they do farther north—Utah, Nevada, and Wyoming.

In short, no part of the United States is without a combination of renewable electricity sources. It is this combination that makes for viable systems, since no one source would be viable alone. Photovoltaic cells, used alone, would require massive electricity storage facilities, at prohibitive expense. Photovoltaic cells, integrated with wind, hydro, and pumped storage facilities constitute a viable system. With other baseload sources like geothermal units, solar ponds, OTEC, biomass cogeneration, and biomass peaking units, a viable system can be visualized in every part of the nation.

Liquid Fuels

Liquid fuel feedstocks are not completely matched to markets, but they are much more evenly distributed than petroleum deposits. If methanol should emerge as the dominant biomass-derived liquid fuel, it could be manufactured in some quantities in every state. Wood, urban trash, crop wastes, agricultural processing wastes, and some energy crops are all potential sources.

Wood supplies are most abundant, per capita, in the Southeast, New England, and the Pacific Northwest. The other sources are available virtually everywhere. Canada is particularly well endowed with wood resources per capita and may well emerge as a methanol exporter.

The volume of liquid fuels transported would be much less than is presently the case. Efficiency measures would have reduced the volume used by two-thirds. Trade between surplus and deficit areas would, as a rough estimate, be a sixth of present shipments, since few areas would be less than half self-sufficient. Virtually all petroleum-based fuels are now transported before use.

Biomass materials that are most likely to be used for methane production are also widely distributed. They are not nearly as abundant as wood fuels,[3] but they will be of critical importance in providing high-temperature industrial heat as well as fuel for cooking. Potential sources are food-processing wastes, animal wastes (feedlots, dairies, poultry farms), and sewage. Much plant matter also can be used. The mere listing suggests wide availability.

Direct Solar Heat

Direct solar heat is also available everywhere. Solar water heaters are found in every state. Recent installations of industrial process heat include such states as South Carolina and Wisconsin. The amount of solar radiation does vary from one area to another, though not as much as might be thought. Annual average insolation in the United States ranges from 400,000 to 860,000 btu's per square foot.[4] This includes radiation received on cloudy or partly cloudy days. Where concentrators are required and only direct sunlight will do, the range widens from about 300,000 to 860,000 btu's per square foot per year. Seasonal variations usually lie within 30 percent of the annual average.

This brief survey, then, confirms that multiple renewable sources of heat, electricity, and liquid fuels are found in all regions of the nation.

State and Regional Studies

Of all the states, California has studied most thoroughly the conservation-efficiency and renewable elements of an energy future. Other state or regional studies deal with parts of the system—conservation potentials, electricity conservation potentials, or renewable sources.

California. California's leading position in energy studies is partly a result of its abundant endowments. More, perhaps, is owed to a perceptive governor supported by an informed and venturesome legislature. The political leadership took an early and sustained interest in energy conservation and renewable energy and created institutions that, so far, have survived in the administration of a new governor whose attitude toward these matters ranges from indifferent to hostile. (The same kind of leadership change occurred at the national level at the same time, and federal programs were dismembered.) The California Energy Commission has gathered and published much valuable information and has had funds to support innovative projects. A similar revolution in thinking about energy matters spread through numerous state agencies, including the Public Utilities Commission. Aggressive utility conservation programs, appliance efficiency standards, a strict state building code, the construction of low-energy state and local government buildings, and local ordinances are slowing the growth in electricity demand. Growth rates between 1 percent and 2 percent are forecast for the rest of the century, even though population growth in the state is well above the national average and per capita income is expected to rise. The major utilities, after some resistance, have embraced renewable electrical energy sources, which have developed rapidly.

The California Energy Commission projects a renewable share in the state's electricity of 50 percent to 60 percent by the year 2000.[5] Hydroelectricity and geothermal sources are well established in the utility industry; both are expanding rapidly. The present wind-electric boom in California is assisted by tax credits, but it still would not have taken place had not the commission done the advance studies showing how large the wind potential was. Though photovoltaic electricity is not yet cost-competitive on grids, California has the largest utility photovoltaic capacity in the world. The Sacramento Municipal Utility District[6] is in the second stage of a 100-megawatt project, and the privately owned utilities have PV generating facilities as well. The commission is helping to finance the Sacramento project, since the gaining of experience with PV systems and their interactions with utility grids is seen as important. The enlargement of the PV market, which these systems provide, is helping the manufacturers to lower production costs. Numerous small-scale hydroelectric projects are under way. The utilities were, until recently, actively seeking out industrial cogenerators, many of which use biomass fuels.

With respect to liquid fuels, the commission sees methanol as the most promising replacement for petroleum products. Methanol is presently made from natural gas, opening a market for gas supplies not connected to pipelines. Coal is another source, but, more important for our purposes, virtually any biomass material can be used as feedstock. The commission is assisting in the simultaneous development of a methanol distribution system and a fleet of vehicles to serve as customers. It is thus addressing the major question of finding a long-run substitute for petroleum in transportation—an issue that is generally neglected as government research programs concentrate on electricity. It foresees a 12 to 13 percent contribution from methanol to the state's transportation energy budget by 2000.[7]

California's approach could be emulated with profit by every national and regional government on the globe. The California experience with electricity supply is most instructive. The following assertions have been made at several points in this book, and in other referenced studies— notably SERI and the Audubon Energy Plan. They are so important that they bear repetition here, so we can examine them along with California's experience.

1. Renewable electricity has numerous sources. Electricity supply is one of the most easily managed aspects of the transition from oil and gas.

2. Efficiency improvements (including cogeneration) and renewable supplies can be brought on line sufficiently rapidly to avoid any new investment in coal or nuclear facilities.

Both of these propositions have been amply confirmed in California. In the late 1970s and early 1980s, state policy, the Public Utilities Commission, and the utilities turned in earnest toward electricity conservation, cogeneration, and renewable electricity. Early progress was slow but steady. Then offerings of nonutility power accelerated, and soon there was far too much in prospect. The Public Utilities Commission is considering less generous terms for purchasing electricity, and the California Energy Commission is reexamining costs to assure that the cheapest sources are brought on line first. The lesson is clear: when conservation opportunities and renewable energy sources (especially electricity) are aggressively sought out, they will appear in abundance. No California utility is considering new coal or nuclear capacity.

The Pacific Northwest.   The Pacific Northwest has been studied with respect to the conservation potential for electricity. The studies represent the outcome of a protracted struggle between traditional and powerful "supply expanders" and those who tried for years to show the economic advantages of the conservation of energy. The Washington Public Power Supply System (WPPSS) nuclear fiasco, which included the largest default on tax-exempt bonds in history, gave added credibility to the "conservation–investment first" concept. This idea has finally worked its way into official documents in the Northwest as it has in California.

The Northwest Power Planning Council prepared a Regional Conservation and Electric Power Plan, which it published in 1983.[8] The plan considers several possible growth rates for population and income, with resulting growth rates in electricity demand ranging from .9 percent to 2.8 percent annually. It then developed a set of conservation-efficiency investments that would meet about half of the incremental demand under the highest-growth scenario. These measures cost about 1.5¢–2¢ per kilowatt hour saved, a result that never fails to appear when the matter is seriously studied.

Since no incremental electricity supply can be had this cheaply, the study has apparently stopped short of the economic margin, a result also consistent with the conclusions of this study. The unspoken conclusion in the council's report is that its conservation measures in other, lower-growth scenarios would displace coal. The fuel and operating costs of coal plants (short-run marginal costs) exceed the estimated costs of kilowatt hours saved. Thus, for the growth rates in population and economic activity that the council thought most likely to occur, electricity demand could well decline.

Should any expansion be warranted, additional hydroelectricity and industrial cogeneration (80 percent biomass) were identified as the next cheapest sources after conservation. Wind-electric generation was not considered, since it was not then competitive nor were its estimated future costs as a mature technology sufficiently firm. The council did expect that windpower, up to 2,000 MW, would be used by 2000, since the region has a large potential.

The council, not wishing to rub salt in wounds, did not stress the point that its alternatives were much cheaper than the now-defunct nuclear plants, nor did it point out that anyone who went to a little trouble could have found this out long ago. It takes no great effort to extract this information from the report.[9]

An Oregon study considered demands and supplies of electricity and other forms of energy in all sectors except transportation.[10] Several renewable sources (hydroelectricity, geothermal power, alcohol fuels, other biomass, windpower, and solar heat) were evaluated. The supply from renewable sources that could be developed by the year 2000 was

ssufficient to meet the entire increment in energy demand foreseen for
that year. Conservation measures expected to be taken by 2000 repre-
sented only a small fraction of the potential identified in the study.
Renewable energy supplies likewise were but a small fraction of the long-
run potential. The results, again, are consistent with the findings of the
present study.

New England.   The New England study[11] notes the observed decline
in per capita energy use through 1980. It does not attempt to estimate
additional possible savings but is focused on the near-term outlook for
renewable energy supplies. Such sources provided 6.3 percent of the
region's primary energy in 1980, with good prospects of providing 13
percent in 1985. The report does not make estimates of renewable energy
use beyond 1985, but it does give some estimates of overall potentials
that are useful for our purposes.

Forest biomass could provide up to .8 quad for energy purposes after
the demands of the forest products industry are satisfied. The hydroelectric potential is about 3,000 megawatts, of which half was already
developed at the time of the study. Another 500 megawatts were in
various stages of construction or planning and were expected to be in use
by 1985. The long-run potential is some .1–.15 quads. (These seem like
modest numbers, but we are looking at a region of 12 million people, not
a nation of 230 million.) Windpower, tidal power, and biomass sources
like food processing wastes and municipal solid wastes are discussed, but
estimates are given only for solid waste (.06 quad) and tidal power (.02
quad).

New England energy consumption per capita was already below the
national average, partly because the most energy-intensive industries are
located elsewhere. Some 3.1 quads were used in 1980. Conservation-
efficiency actions like those discussed in chapter 2 would bring this
figure down to .8–1.1 quad. Since more than 1 quad is available from
wood, solid waste, hydroelectric, and tidal power, and still more is available from wind, photovoltaics, other biomass, and direct solar heat,
renewable sources could easily reach 1.5 quads. New England could
become an exporter of energy, probably in the form of biomass-derived
liquid fuels.

The New England report does not go beyond listing renewable sources
and long-run estimates for a few of them. The remaining conclusions are
mine.

New York City.   New York City is a special case of particular interest.

It permits us to estimate the conservation potential and the local renewable energy potential and, in addition, to address another important issue. We have noted earlier the allegations of a fatal flaw in solar–renewable energy systems—the inability to deal with highly concentrated demands.

We have, thanks to the New York City Energy Office, a particularly detailed analysis of energy use in New York City in 1979.[12] The study looks at the near-term conservation potential as well. We shall review that study and then allow for additional gains in energy efficiency that could have been achieved if conservation-efficiency measures were taken as far as in chapter 2.

New York City also is a potential producer of energy. We shall identify potential outputs of renewable energy in the city; they will amount to about one-fourth of its energy consumption once full efficiency gains have been achieved. The remaining three-fourths can be supplied mostly from the state of New York or perhaps New England.

These contrasting views as to the viability of renewable energy can be compared best by quoting one of the critics. Petr Beckmann's "Solar Energy and Other 'Alternative' Energy Sources"[13] discusses the industrial revolution and the pivotal role played by coal in allowing the highly concentrated use of energy. In turning to solar energy, he says:

> Conversely, the density of energy consumption, that is, energy consumed per square meter in the urban areas of the world—which host an ever increasing fraction of its population—cannot be supplied by an energy source as dilute as solar power. . . .
>
> For example, the average density of power consumption in the urban areas of West Germany is 7.5 W/m^2; in the urban areas of India the average is 12 W/m^2. . . . Careful studies have shown that urban settlements with population densities of more than 1,000 persons/km^2 (London: 1,100; Rhine-Ruhr area: 1,280) cannot exist without some sort of centralized power, quite apart from any cost considerations.[14]

New York City has a population density of about 8,000 people per square kilometer and uses energy at a rate six times greater than the one he cites for urban West Germany (his sources evidently include suburban and surrounding nonurban areas to get his density figures). We shall see, as we consider New York, that Beckmann is partly right, but that his conclusions do not necessarily follow.

The New York study, *Energy Consumption in New York,*[15] estimates that New York City used about 1.3 quads in 1979. This is considerably less than the national average, but at considerably more cost. Residential use per capita is lower since so many New Yorkers live in apartments, which use less space heat per square foot than do detached structures. Transportation energy use per capita is also lower when only ground transportation is considered. The inclusion of fuel sales at two major airports (and for ocean shipping) inflates the transportation portion of the 1.3 quads. Industry use per capita also is less than the national average since highly energy-intensive industries are underrepresented in New York City.

The energy office estimates that 1979 energy consumption could be reduced about 20 percent by low-cost or no-cost actions. These, if implemented, would reduce energy consumption to about 1 quad (all figures are in primary terms—before deductions for conversion losses as in generating electricity). In most cases, the authors of the study took only about 50 percent of the potential savings they identified;[16] this was their estimate of the near-term adoption of the measures they considered. This implies that, if the low-cost or no-cost measures were fully implemented, energy use could have been about .8 quads (or 790 trillion btu's. In considering smaller areas than the nation, it will be useful to scale down our energy units. 1,000 trillion btu's equals 1 quad). As one gets into the details of the estimates, more than half of the projected savings were in fact included in the office's figure, so that a reduction to about 845 trillion btu's would be more accurate.

This is as far as the study goes, which is quite far, actually—the implementation of these measures would lead to a gradual decline of 35 percent in energy use in New York City.

Suppose now that we take the very detailed end-use categories for energy consumption, and then apply our own long-run potential reductions. Energy use for the 1979 output in the city falls to about 450 trillion btu's. The largest reductions are in space heat (insulation, infiltration reduction, solar gain) lighting efficiency in commercial buildings, vehicle fuel efficiency, and aircraft fuel efficiency.[17] Industrial use declines as electric motor efficiency is increased and process heat is better used. Space heat in industrial structures, a small fraction of industrial energy use nationally, is more important in New York. Reductions in this category are made in ways similar to techniques used in residential and commercial buildings.

It is this major reduction in energy use through investments in conservation-efficiency measures that undergirds our whole analysis. Its neglect by Beckmann and similar critics is the first major difference in approach. One must be cautious about declaring what is impossible and what is possible. Parameters that one takes for granted as fixed (in this case, energy use per capita) sometimes turn out to be the most important variables.

Next we inquire into prospects for energy production in the city. Sources of renewable energy that are already beginning to be used include:

Combustion of solid wastes for heat, steam, and electricity.

Methane from landfills (eventually exhaustible, of course, if landfilling ceases in favor of combustion of wastes).

Methane from sewage.

Photovoltaic electricity (where space is at a premium, from rooftops).

Direct solar water and space heat.

Direct solar industrial process heat.

Other possibilities include the production of liquid fuels from urban biomass sources and small amounts of wind-generated electricity.

One can make some estimates of the energy available in New York City from these sources. The present volume of solid waste is about 22,000 tons per day, or 8,000,000 tons per year with a gross energy value of about 72 trillion btu's. The city is presently considering a number of trash-to-electricity plants, prompted mainly by a concern that landfills are rapidly approaching their capacities. The main environmental concern seems to be that some garbage components, when burned, produce dioxin. Recycling, especially of paper, would reduce the volume of combustible waste. In the long run, it might seem more desirable to use organic components of solid waste to make liquid fuels rather than electricity. Nevertheless, the presently available resource is about 72 trillion btu's.[18]

Methane from past landfills is already being produced in New York City.[19] Full exploitation would yield some 5 trillion btu's.

One city sewage treatment plant in Brooklyn has arranged to provide methane gas to Starrett City[20] for electricity generation. The heat, usually wasted, is to be returned to the plant for its heating needs, including keeping the digester tanks warm. Similar arrangements at other plants could bring a total of about 1 trillion btu's.

The use of photovoltaic cells to produce electricity need not be constrained by spaces to put them. New York City has an estimated 700 million square feet of residential rooftop space, even though half the population lives in high-rise structures.[21] Commercial and industrial rooftops provide more potential sites. One need not go through a detailed analysis of shading and other determinants of site suitability to estimate that at least 10 to 15 percent of this area might be used. Some 13–18 trillion btu's of electricity could be provided.[22]

Providing 50 percent of water heating for the 50 percent of New Yorkers not living in high-rise units would provide some 8.5 trillion btu's. Suitable commercial and government buildings could bring this to 12 trillion, as could modest use of active solar space heat or some use of collectors on high-rise buildings.

Industrial process heat is said to require large amounts of space per unit of delivered energy. Yet 2–4 trillion btu's could be had in New York by using 10–20 million square feet of rooftop area or parking lots (some 240–480 acres, a not impossible task in New York's 365 square miles).

These sources sum to 90–96 trillion btu's—about one-fifth of New York's efficiency-reduced energy demands. If solid waste is used to generate electricity, then, New York City would be about 35 percent self-sufficient in electricity. About half the low-temperature heat need would be met, but only about 20 percent of high-temperature heat demands. All liquid fuels for transport (37 percent of energy demand) would have to be imported, as they are now. These would be available from biomass, mostly in New York and New England.

The other 65 percent of the city's electricity demand would be available from hydroelectric and wind-generating facilities in the state of New York.[23] Some of the low-temperature heat deficit could be met if the garbage were burned in units that cogenerated electricity and steam. This would raise the self-sufficiency ratio by a few more percentage points. All of this leaves some 28 trillion btu's of high-temperature heat to be provided by imported fuels as the most difficult part of the energy supply problem. This also could be supplied from nearby areas from biomass sources.

Critics like Beckmann, quoted above, conclude too readily that renewable energy cannot support concentrated economic activity. It may be true that they cannot be supported entirely by small-scale facilities entirely in the city limits (though the percentage of self-sufficiency could readily have been pushed up with more direct solar use). But electricity and biomass-based liquid fuels can be transported cheaply (as they are now). In fact, with this much conservation/efficiency and with so much local production, imports would be about one-fourth their present level.

Centralized, large-scale facilities would exist at large hydroelectric facilities and, to a lesser degree at trash-burning plants. Methanol, ethanol, or other liquid fuels would probably be produced at numerous small- or medium-scale facilities placed out where the raw materials are.

Issues of renewable energy sufficiency, therefore, should not be confused with the issue of centralization-decentralization. To be sure, many solar-renewable advocates have not been clear on that point either. The system we envisage, or any others that would develop which we cannot now envisage, would probably be much less dependent on centralized facilities than is the case at present.

The selection of these four cases for a brief survey of local-regional issues in the United States has been governed in part by the availability of studies. They do, however, include 50 million people—over one-fifth the nation's population. To be sure, areas without wind or hydro resources, or with a high concentration of energy-intensive industry are underrepresented. Since national totals are adequate, at worst we are left with a problem of transporting electricity, liquid fuels, and methane in much smaller quantities than we do now. Since no metropolitan area is likely to be less self-sufficient than New York, an upper limit on energy transport of three-fourths for the least well-endowed areas seems appropriate. If the average degree of self-sufficiency is no less than one-half, and if the nation is using one-third to one-fourth the 1980 level of energy inputs, then energy transport would be one-sixth to one-eighth the present level.

To go beyond this brief analysis would imply more precision than exists. The purpose of this chapter has been to flesh out the analysis of chapters 2 and 3 by looking more specifically and less abstractly at parts of the nation. We also have dealt with the allegation that renewable energy sources cannot supply concentrated populations. We have shown this not to be the case even with presently available technologies. There is so much flexibility in the renewable energy supply system that we cannot even predict which configuration is most likely to emerge.

5

What of Coal, Synfuels, and Nuclear Power?

The last four chapters have dealt at length with issues of energy conservation, energy efficiency, and renewable energy. They have barely mentioned coal, synthetic fuels, and nuclear power. Since this study stresses energy efficiency and renewable energy, it might seem appropriate to omit any reference to other sources. Yet a discussion of other sources cannot be avoided.

Economics is concerned with choices among alternatives. Our whole approach has been one of comparing costs and returns of energy conservation investments, renewable energy investments, and investments in new conventional energy supplies. Over the next several decades, as oil and gas production peak and then begin to decline, the relevant comparisons are increasingly with coal and nuclear power. Moreover, governments, international institutions, and the energy supply industries have all viewed the future as one of increased reliance on these sources. These views, so widely held, must now be examined in comparison with the ideas expressed in this and numerous other studies that stress conservation-renewable energy.

Since energy analysts with a coal-synfuels-nuclear orientation have notions of an energy future so different from mine, an objective comparison is difficult. I shall attempt to treat these views, insofar as possible, on their own terms. I shall first lay out, from representative sources, ideas developed in the mid-1970s as to what the world would be like in 1985 and 2000 as far as energy supplies and demands were concerned. We are now in 1986, and therefore in a position to apply all the wisdom of hindsight to the earlier projections. We also can now comment on what the year 2000 seems to hold in prospect, as viewed from 1986.

Projections from the 1970s

In world energy matters the primary organization is the International Energy Agency. It was formed within the Organization for Economic Cooperation and Development in 1974 as a Western response to the first oil "shock." Its 1977 projections are quite typical of national and international projections made at that time. In looking ahead at probable energy demands and supplies in the OECD nations (the United States, the countries of Western Europe, Japan, Canada, Australia, New Zealand), it foresaw energy demands rising from about 150 quads in 1973 to 213 quads in 1985 and 365 quads in the year 2000.[1] Similar projections for total energy demands were made by the Workshop on Alternate Energy Strategies (WAES), sponsored by the Massachusetts Institute of Technology. The WAES projections, for regions of the world and for the entire world, excluding communist nations, were made by a group of energy experts from fifteen nations—twelve OECD members and three oil suppliers. WAES projections are somewhat larger than IEA's since they include developing nations as well as OECD nations.[2] With this adjustment, they are quite consistent. U.S. projections made about the same time showed similar growth in energy demand—from 74 quads in 1973 to about 100 quads in 1985 and 140 to 160 quads in the year 2000.[3] (These projections are shown in figures 5 and 6.)

All of these projections contemplated greatly increased use of coal and nuclear power, and, toward the end of the period, some "synfuels"— liquid and gas fuels extracted from coal, oil shales, and tar sands. The IEA study expected coal use to double by the year 2000. Some four hundred to five hundred large nuclear plants were expected to be operating in the OECD nations by 1985, and two thousand or so by 2000. The WAES study also saw coal use rising by 50 percent in 1985, and doubling or tripling by the year 2000. Its nuclear projections for 1985 were only slightly smaller than the IEA's, and its projected range of facilities for the year 2000 was 900 to 1,800 large nuclear plants.

U.S. plans called for doubling coal output by 1985, and tripling or quadrupling it by the year 2000. Nuclear capacity was expected to be two hundred to three hundred large plants in 1985 and seven hundred to a

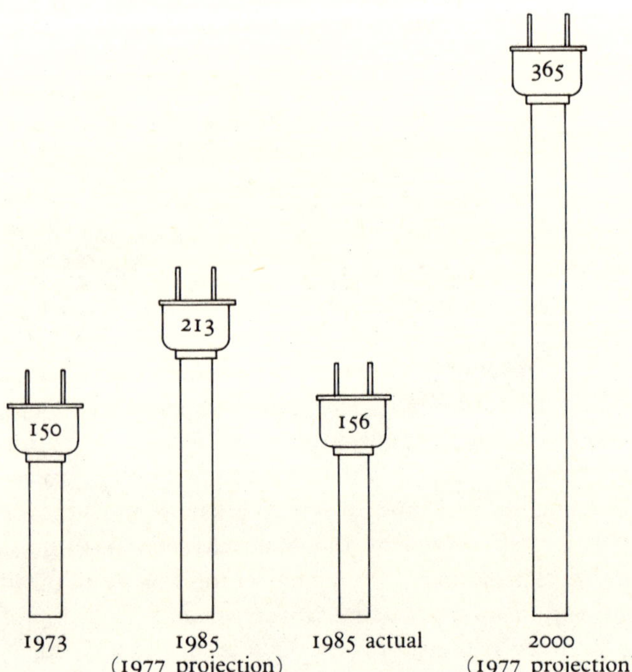

Figure 5 Energy Demands in OECD Nations 1973, 1985 (as projected in 1977), 1985 Actual, and 2000 (as projected in 1977) (quads). Sources: Projections from *World Energy Outlook*, 1977 Actual; OECD, including preliminary 1985 data.

thousand plants in the year 2000. A synfuels output of 2–4 quads (1–2 million barrels a day) was expected by 2000.

These expectations for coal, synfuels, and nuclear outputs in the United States are also shown in figure 6. As that figure indicates, it also was expected that oil and gas production would rise and remain at or near physical limits. Increments on the output of total energy, however, would depend heavily on coal and nuclear power.

The expected evolution of the nuclear power industry is particularly important. The rapid expansion foreseen would require an expansion of uranium mining, processing, and fuel enrichment. The projected scarcity of uranium would require the reprocessing of spent fuel to remove still usable uranium and plutonium.[4] Breeder reactors would be developed, since they could convert a more abundant but otherwise useless form of uranium into more plutonium. Waste handling and long-term waste disposal procedures would have to be developed.

That is to say, a nuclear reactor that produces electricity is only one component of a large and complex system of fuel supply, reprocessing of spent fuel (to remove plutonium for breeder reactors and still usable uranium), further handling of highly radioactive nuclear wastes, and permanent disposal of such wastes (the "back end" of the fuel cycle).[5] Procedures also would have to be developed for the disposition ("decommissioning") of reactors themselves after their useful lives of thirty to forty years.

It is significant to note that the world's commercial nuclear power

Figure 6   U.S. Energy Demand, 1973; 1985 as projected in 1975; 1985 Actual; and 2000, as projected in 1975 (quads). Sources: Energy Research and Development Administration, *A National Plan for Energy Research, Development and Demonstration: ERDA 48* (Washington, D.C., 1975); *Monthly Energy Review*, February 1986.

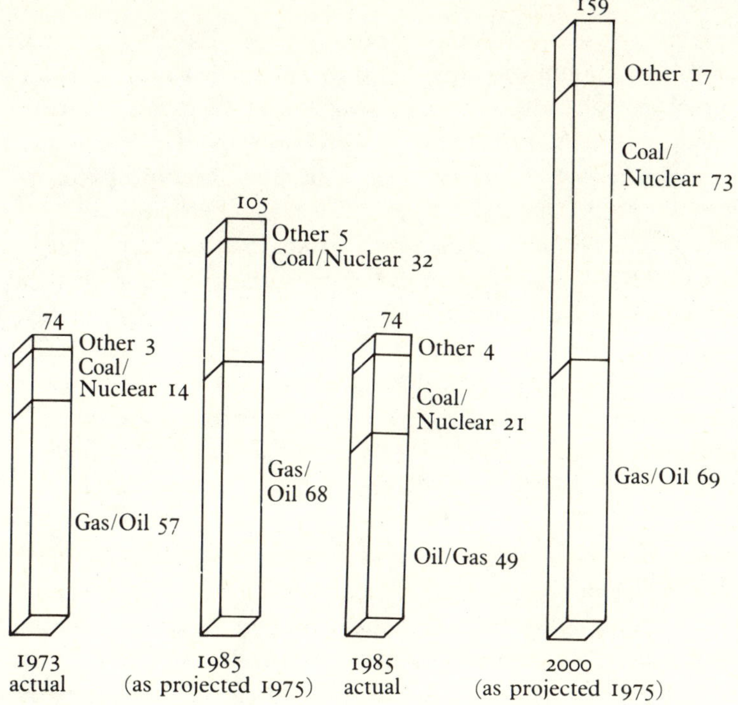

industry was launched with only part of the system fully in place. The "back end" of the fuel cycle (reprocessing, breeders, and waste disposal) as well as plant decommissioning were by no means technologically mature. Reactors themselves were scaled up rapidly before there existed a long history of construction and operating experience with smaller reactors. Confidence that the remaining technical developments could be accomplished was no doubt well placed; so much had already been successfully accomplished. The only point here is that there is a difference between confidence, on the one hand, and actual operating experience, on the other. This, of course, is true of any technology.

1986 Outcomes

From the vantage point of 1986, one can now compare projections with outcomes, and projections for total energy, coal, and nuclear power with actual figures. That they differ considerably is not meant to reflect on the forecasters. The important thing is to see what can be learned from the comparisons.

Total energy use in the United States for 1985 was about 74 quads—similar to the amount of energy used in 1973. That may be compared with projections already mentioned—100 to 105 quads. OECD use in 1984 was about 155 quads. This figure was probably about 156–158 quads in 1985—a sum nearly 30 percent below the 213 quads projected (as recently as 1977) for 1985.

Readers will, perhaps, immediately notice that economic outputs in 1985 are well below those contemplated when the projections were made. Unemployment is still high; U.S. output is about 5 percent below a "high employment" level,[6] and output in Europe is even further below. When one roughly corrects for these differences, 1985 energy use might have been about 165 quads,[7] still nearly 25 percent below the 1977 projections. The difference is about equal to Western Europe's total energy consumption. In other words, this is not a minor forecasting error.

The shortfall in total energy consumption has, naturally, been reflected in petroleum markets, so that petroleum prices have precipitously fallen. It is important to note that the extra petroleum that would have been demanded in a "high employment" world is still well below the world's 1985 oil production capacity.[8] Electricity demand growth also has been well below 1977 forecasts, a matter of particular significance for coal and nuclear power use.

From 1973 to 1984 U.S. coal use did indeed rise about one-third, as compared with the doubling expected in 1977. OECD coal use, which, of course, includes U.S. use, rose 20 percent[9] over the same period, as contrasted with a projected increase of 50 percent.

OECD 1985 nuclear capacity was about 215,000 megawatts, some 30 to 50 percent below the estimates given above. About half of the shortfall is in the United States,[10] the other half abroad. This is an important point; American nuclear interests decry the lag in the U.S. industry and tell us how well the rest of the world is doing. Yet all OECD nations except France are well below 1977 forecasts. (France, a highly significant case, is treated in an appendix to this chapter.)

To summarize, 1985 energy use turned out quite differently from 1977 expectations.[11] Total energy use in the United States and other developed nations is virtually the same as in 1973, not up by 30 to 40 percent. Coal use has grown, but at rates far below projections. Nuclear power output also is far below 1977 expectations, even though nearly all plants expected to be operating in 1985 were under way in 1977 or at least on order.

The Year 2000 Revisited

Forecasting is hazardous; that is not news. Anything now said about the year 2000 will probably be equally far off the mark when the actual figures are in. Still, it is instructive to see how views of energy demand likely to exist in that year have changed over the period from 1977 to 1985. Figures 7 and 8 give some official forecasts of energy demand for the year 2000 in the OECD nations and the United States as they were made at various times. Since 1977, projections of energy use in the year 2000 have generally fallen by about half. These are official forecasts; the energy demand studies cited in chapter 2 give much lower estimates.

Nuclear Issues

Of particular interest with regard to the nuclear fuel cycle are developments in reprocessing capabilities. Stated world capacity is 610 tons of used reactor fuel, enough to accommodate 2 to 5 percent of spent fuels

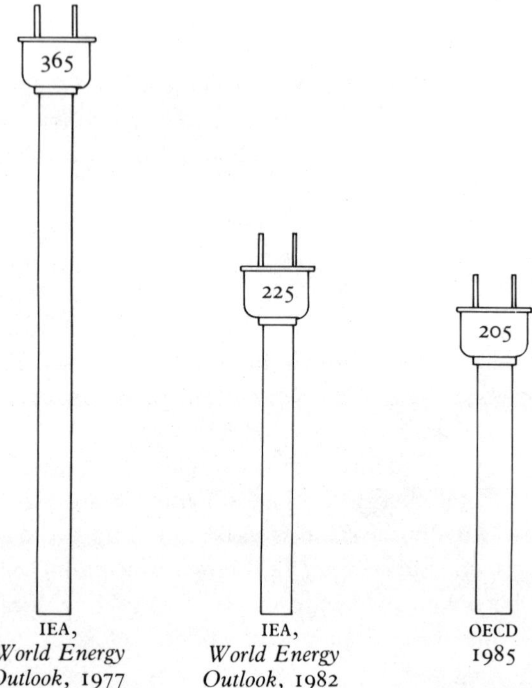

Figure 7   Energy Demand in OECD Nations, Year 2000 Estimated in 1977, 1982, and 1985 (quads).

produced annually by reactors now operating.[12] No commercial facilities operate in the United States, and none have done so since the period from 1966 to 1972.[13] By way of comparison, the WAES study summarized national plans for reprocessing capacity; there was expected to be enough to handle five thousand tons of spent fuel annually in 1985.[14] This is a most significant development, often overlooked in discussions of nuclear issues.[15]

The development of breeder reactors also has lagged far behind 1977 expectations. The United States has abandoned its plans to build a demonstration reactor. Indeed, it has abandoned the technology altogether, or deferred it for several decades, depending on one's point of view. Breeder programs continue, but far behind schedule, in Japan, the United Kingdom, and West Germany.[16] France, the world leader, opened its fully commercial, large-scale breeder reactor late in 1985. It was planning to follow this facility with four to six more, but they have been canceled.[17]

No nation has yet decommissioned a reactor of commercial size; that process is beginning in the United States. Costs, therefore, are estimates, not yet based on experience.[18] No permanent waste disposal has yet taken place; costs, therefore, of this component of the fuel cycle are also estimates. Scientists in the field generally agree that there are no insuperable technical obstacles.[19] Still, the fact is that waste disposal has not yet been carried out. There is no experience on which to base cost. In summary, the back end of the fuel cycle does not exist in the United States, and the situation abroad is not altogether different. Reprocessing capacity de facto exists for only 2 to 5 percent of spent fuel when one looks at actual operating rates of reprocessing facilities. The other 95 to 98 percent of wastes accumulate in temporary storage abroad, just as they do in the United States. Permanent waste disposal exists only in

Figure 8 U.S. Energy Demand, Year 2000 Projections made at various dates (quads).

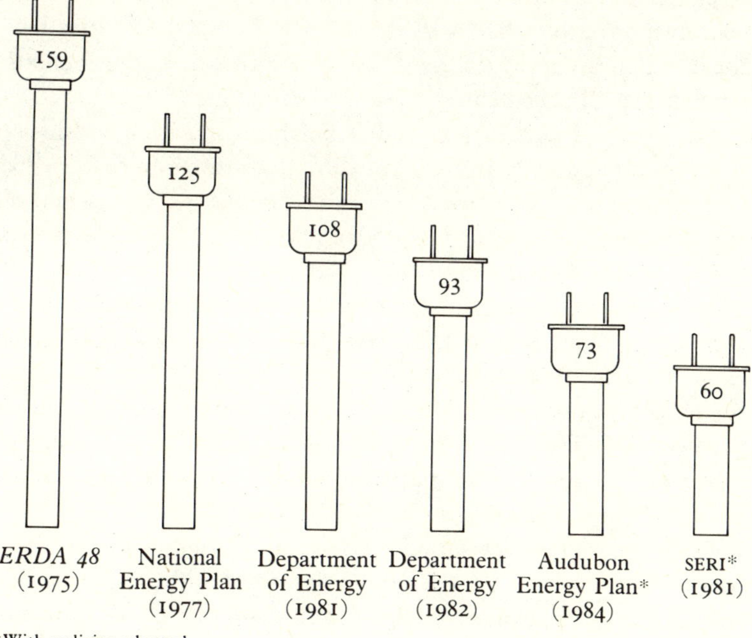

ERDA 48
(1975)

National
Energy Plan
(1977)

Department
of Energy
(1981)

Department
of Energy
(1982)

Audubon
Energy Plan*
(1984)

SERI*
(1981)

*With policies adopted.

plans and concepts, actual disposal has not been carried out. Breeder programs are far behind expectations for 1985.

Some Tentative Conclusions

What lessons may be learned from these observations? What do they have to do with energy efficiency and with renewable energy? The most important lesson is that the energy conservation-efficiency potential is much larger than it was thought to be in the 1970s. U.S. productive output has grown by a third since 1973, while energy use has stayed about the same. Output has grown even more in Japan, again with little increase in energy use. Europe has had a smaller gain in output, but a decline in overall energy use.[20] Investigations into the conservation potential show that much of it already existed at 1973 price levels; two oil price increases have focused attention on opportunities and have, of course, enlarged the scope of efficiency-enhancing investments.

As an interesting observation, it is the centrally planned national economies that have not shown much flexibility in energy-efficiency matters. The Soviet Union and other nations in Eastern Europe use much more energy per unit of output than does the West.[21] Energy-GNP ratios have not changed much since 1973; a great deal of the growth in world energy use has taken place in the centrally planned economies.

A second observation is that there has not been much conversion to coal in the industrial sector. In the United States, growth in coal use has been for the generation of electricity.[22] Coal is cheaper per btu than is oil or natural gas, but the equipment to burn it cleanly is more expensive, especially for small installations.

A third observation is that nuclear power has not turned out to be as cheap as it was expected to be in the early 1970s. This is well known to be the case in the United States. It also is true abroad.[23] Real electricity prices have risen in the United States, in Japan, and in Europe. This is partly because of the increase in world oil prices (oil generated 25 percent of OECD electricity in 1973)[24] and partly because the expected cost reductions in nuclear power did not materialize.

All of the changes in energy supplies and demands in the United States for the period from 1973 to 1984 can be summarized in the following few sentences. If energy use had grown at the same rate as the GNP, it would have been about 98 quads—rather near the forecasts for 1985. The actual figure, including biomass and other renewable sources not in official figures, was below 77 quads. The 21-quad "conservation-efficiency bonus" dwarfs anything on the supply side. For the rest, oil and gas use dropped by 8 quads, coal use increased by 4.2 quads, nuclear power increased by 2.6 quads, and renewable energy increased by 3 quads. Energy "supplied" by conservation, even after adjustments for various other factors, was the leader by an overwhelming margin.

A 1980 Coal-Synfuels Nuclear Economy?

We now proceed further to examine the prospects for coal, synthetic fuels, and nuclear power by asking exactly the same questions of them that we asked earlier of conservation and renewable energy. Again, this is a kind of thought experiment.

Could coal and nuclear power (given adequate time for development) have supplied the energy demands of a U.S. 1980 GNP? Could they have done so at costs no greater than the long-run marginal supply prices given in chapter 2? If further technological developments would be needed, or cost reductions, was there reputable engineering opinion that those changes might be achieved in a decade? Could the economy, using only coal, synfuels, and nuclear power, grow from 1980 levels?

Energy use in 1980 in the United States, it may be recalled, was 76 quads. Could that have been supplied? If so, could one find the right mix of electricity, industrial and other heat, and liquid fuels for transportation? That is, would the sources match the end uses? This is not idle speculation. Nearly all of the high-energy growth scenarios that formed the basis of public policy in the United States contemplated *increments* in coal, synfuels, and nuclear power of this magnitude by the year 2000. So these questions are more than matters of symmetry with the chapters on conservation-renewable energy.

Coal and nuclear power supplied about one-fourth of the nation's energy in 1980. Could they have supplied four times as much energy? As to technical capability, no doubt the answer is yes. As to cost, the criteria used in chapter 2 could not be met. Long-run marginal costs in the coal industry are not thought to rise very much, but they do, no doubt, rise to some degree. A fourfold expansion, then, would raise coal costs. A nuclear industry four times as large as 1980's would probably be more costly, too. An industry that large, and growing would, of necessity, bring reprocessing and breeders back into the calculation of costs. These costs are not known but are almost certainly higher than the costs of a system without them. Cost-lowering technical development and technical maturity of breeders and reprocessing would have to be established, as would an assessment of the probabilities of that happening within a decade.

Next, we examine the matching of sources to end uses in a coal-nuclear energy supply system for the 1980 GNP. First, consider electricity. These two sources already provided over 60 percent of the nation's electricity in 1980. Since there is no reason to dispense with the hydroelectric supply system (a kind of symmetry for keeping metallurgical coal in a renewable energy future), the electricity supply should be rather easily managed. There would be peak demand problems since nuclear power is entirely baseload power, and coal-fired capacity is better suited to baseload capacity than to anything else. The hydroelectric and pumped storage systems would accommodate most of the demand pattern requirements.[25]

Demands for heat are less easily accommodated. Some industrial heat is already provided by coal. All of it would have to be, unless electricity or fossil synfuels were to be used. Pollution regulations, which represent an attempt to internalize costs previously borne by others, raise costs and require new technologies in the burning of coal to provide industrial heat. That is the main reason that oil and natural gas continue, twelve years after the first oil price shock, to provide over 85 percent of fuel for industry. Coal as a fuel has long been cheaper per btu—it is more expensive to handle and to burn cleanly. There are economies of scale in pollutant removals so that with present technologies, coal use becomes increasingly expensive per unit of energy as the application gets smaller and smaller.

New combustion technologies (fluidized bed combustion, for example) are being developed and give promise of being cost-competitive in the next few years. Gas also can be extracted from coal and then burned cleanly in the factory. Other heat demands (space heating or cooking, for example) are already met in part by electricity. It is much more expensive per btu than heat from fuel. The cost disadvantage for space heating can be partially overcome with heat pumps. This exacerbates peaks and would probably require some gas combustion (coal-derived gas) to meet peaks. Replacing all fuel for space heat in 1980's buildings would have required a 50 percent expansion in the quantity of electricity supplied. Since all of the additional electricity would be used in less than half the year, the industry would have to double in size. Peak demands would add some additional peaking capacity. This consideration further pushes up costs.

Transportation presents a different set of end-use energy requirements. More of the railroad system could be electrified, since electricity is the kind of energy most easily provided in a coal-nuclear regime. For surface vehicles, suitable batteries must be developed if electricity is to be used very extensively. Batteries must be relatively cheap, light relative to their storage capacities, long-lived, and capable of many cycles of deep discharge-recharge. For surface vehicles not powered by electricity and, in all cases, aircraft, liquid fuels must be available. These would come from coal, oil shale, and tar sands. Large quantities of such fuels would have to be available to meet 1980 demands—some 20 quads—nearly 10 million barrels per day.

It is not known whether batteries with the requisite characteristics can be developed, though some progress has been made.[26] Synfuels have been and can be produced in small quantities. Ten million barrels a day is another matter. Costs are not known. There are estimates, but there are no experience-based costs in the United States for liquid synthetic fuels from fossil sources. Estimates range from $40 to $100 per barrel,[27] compared with the long-run marginal supply price for crude oil of $40.

The issues, then, come back to those of costs and the probabilities of the required technical and cost-lowering developments taking place within the decade. The requirements for meeting the 1980 energy demands with coal, synfuels, and nuclear power (allowing 3 quads of hydroelectric power to continue) may be summarized as follows: (1) commercially successful coal combustion technologies that meet air pollution standards are usable at various scales and cost no more, for heat supplied, than oil or gas at $5 to $6 per million btu's; (2) workable processes for extracting liquid fuels in large quantities from fossil sources, at costs in the $40 per barrel range; (3) development of vehicle batteries that are light, compact, cheap, powerful, and durable; (4) workable and economical nuclear fuel reprocessing; (5) breeder reactors that produce electricity at 10¢ per kilowatt hour; (6) establishment of plant decommissioning and waste disposal procedures at or below presently estimated costs.

This listing suggests that a coal–synfuels–nuclear energy supply system cannot compete, at an energy level of 76 quads, with conservation-efficiency investments, and such a system may not even be possible. All costs are equal to or greater than the benchmark costs cited in chapter 2. The conservation-efficiency technology already is known, not a "wish list" of developments. Conservation-efficiency costs are less than the benchmark costs. The barriers to more rapid achievement of efficiency gains are not economic or technical. They have to do with the diffusion of information to large numbers of people, the provision, in some cases, of investment funds, and the overcoming of inertia. The barriers to new energy supply are economic, and, in some cases, technical. This would be equally true for conventional and renewable supply, especially at the 76-quad level. One may conclude that a 76-quad coal-nuclear-synfuels economy is very costly and may not be technically possible.

A 1980 Energy-Efficient Coal-Synfuels-Nuclear Economy

Following the method of examining issues developed in chapter 2, we next ask whether an economy already made energy-efficient and now demanding 20 to 25 quads could be supplied with a coal-synfuels-nuclear supply system.

In aggregate terms, the answer is clearly yes. Coal and nuclear power already supply 21 quads. Electricity demands would be covered several times over. Industrial heat demands, much reduced, would be more easily met, as would remaining, much-reduced demands for heat in other sectors. Transportation demands of 5 to 6 quads still would require liquid fuels, subject to a reduction of one-third to one-half if suitable batteries were available.

Nuclear power at this much-reduced level would no longer require reprocessing or breeders. Otherwise, the list of required technical developments and cost reductions would be exactly the same as the one just given.

Since both coal-synfuels-nuclear and renewable energy futures favor much conservation, we now can make a direct comparison of the two supply systems at the 20- to 25-quad level. Both would deliver about 8 quads of electricity, 6–9 quads of heat, and 5–6 quads of transportation energy, now nearly all liquid fuel.

A Comparison: Conservation-Renewable Energy, or
Coal-Synfuels-Nuclear Energy

Electricity.   In the one case, electricity would be supplied by coal and nuclear plants (keeping the hydroelectric system). In the renewable energy case, electricity would be supplied by a somewhat enlarged hydroelectric system, geothermal, photovoltaic, wind, other solar thermal and biomass cogeneration sources. The requisite technologies are all in place on the renewable energy side. At issue is the pace and probability of cost reductions in wind, photovoltaic, and solar thermal systems. Biomass cogeneration already is competitive in the forest products industry. Also at issue in the biomass area is the pace of cost reduction for other technologies, particularly methane.

On the coal-nuclear side, the technologies also are mostly in place. Cost reduction may be needed to assure that nuclear electricity stays at or below the specified benchmark cost of 10–12¢ per kwh. Technologies for plant decommissioning and permanent disposal of radioactive wastes would have to be demonstrated at costs equal to or less than current estimates.[28]

As it turns out, we do have some experience with direct competition between these two options—actually between the three options of conventional electric supply, renewable electric supply, and more conservation. Potentially, these competitions exist everywhere. Actually, they exist only when the institutions have been developed so that competition on more or less equal footing takes place—where the rules are more or less fair and there are firms to deliver on all three sides of the market.

California is the case in point. All new electricity supply (after the opening of the last nuclear plants still under construction) is expected to come from conservation, renewable sources, and cogeneration. The latter is, of course, a conservation technology. There is such a potential surplus of electricity from the conservation-renewable side that demands for decades to come can be met.[29] Had these sources been identified earlier, the last two or three nuclear plants in California would not have been built.

Heat.   In units too small for coal, the choice is between electricity, fossil synfuels, biomass fuels, or solar heat. For low-temperature uses like water heating, solar collectors are already the least costly solution in this set of choices. When the solar fraction is optimized on electricity at marginal supply prices, not present rates, it gets larger still (more storage pays). Residential space heat nearly vanishes in an energy-efficient economy. Commercial space heat is so integrated with lighting and air-conditioning that electricity is more likely to be used despite its high cost per btu. For other residential-commercial heat uses like cooking or clothes drying, electricity or biomass-based methane are sources in the renewable energy scheme. Electricity or coal-based gas would be the competitor from the other system. Coal-gas ("town gas") was widely used in the past before the age of coast-to-coast gas pipelines from abundant gas wells. Pollution control considerations would now rule out the same installations.

Industrial heat is a coal-biomass-solar contest, given the high cost of electricity for most industrial heat-using processes. At the moment, coal and wood have about the same share of industrial heat, with much larger portions being supplied by oil and, especially, natural gas. An energy-efficient industrial sector producing the 1980 output would use about 4–5 quads of heat. That is technically within the capacity of wood and biomass-methane. These, or coal, or solar heat, would divide the market depending on relative rates of development and reductions in costs.

Transportation—liquid fuels. The move toward an energy-efficient economy would so reduce these demands as to stretch out oil resources for a very long time. Ultimately, fossil synfuels, biomass fuels, and battery-electric systems would compete for this market. There is regrettably little operating experience with fossil synfuels or with biomass-based methanol. Ethanol from grain now costs $1.00 to $1.30 per gallon ($12 to $16 per million btu's) in the United States, while gasoline at long-run supply prices from chapter 2 is $1.00 to $1.10 per gallon before taxes ($12 to $13 per million btu's). The nation badly needs several small operating facilities for each of the liquid fuel sources—shale, tar sands, coal, and biomass. There would then be a source of engineering design refinements and a much firmer basis for estimating costs. The Department of Energy estimates that biomass-derived methanol would now cost over $13 per million btu's (85¢ per gallon). Its target for the year 2000 is 50¢ per gallon (today's prices), about what methanol from natural gas costs. Those who follow biomass energy research also expect ethanol from wood and other biomass sources to become available and to be cheaper than grain-based ethanol.[30] Tax incentives favor ethanol, while excess world capacity in fossil-based methanol discourages research on biomass conversion processes. As chapter 2 pointed out, methanol, rather than more complicated synthetic gasolines, may well be the preferred fuel of the future. It can be made from natural gas, coal, or biomass feedstocks. If enough vehicles are equipped to use it so that a sizable market exists, it seems to be ideally suited to a smooth transition. This exercise of imagining a 1980 economy powered by coal, synfuel, and nuclear power, or one first made efficient and then so powered, leads to the following conclusions:

1. Coal, nuclear power, and synthetic fuels cannot compete with energy-conservation investments. It is not even clear that they could provide 76 quads from a purely technical point of view.

2. For an energy-efficient 1980 economy using 20–25 quads, several technical developments and cost reductions are still required.

3. At this point we move into opinions and judgment, not facts.

Which set of developments seems more likely—that wind and photovoltaic devices will decline in cost, or that nuclear plant costs can be reduced, and decommissioning-waste disposal accomplished at costs below current estimates? Will industrial coal-burning technologies that are nonpolluting and cheap develop more rapidly than solar-industrial equipment or biomass methane technologies? Will methanol, ethanol, or fossil-synfuels emerge as supplements to, or successors to, gasoline?

Opinions vary. Mine are clear: efficiency investments outperform new energy supply whether it be conventional or renewable. Looking at the necessary technology and cost developments, the renewable sources are more likely to succeed.

These conclusions may be reached on cost and technical criteria alone. Coal and nuclear futures carry other burdens that have not even been mentioned. Both have characteristics that have been called "irreducible risks."

Fossil fuels, especially coal, cannot be burned without increasing the carbon dioxide content of the atmosphere and thus threatening a climate change. Nuclear power, even if all other problems were solved, cannot be used without the risk of weapons proliferation.

When these considerations are introduced, it is difficult to see how anyone, once informed of the alternatives, would fail to choose them.

Perhaps all parties could agree to the following set of rules to see which energy future is, in fact, technically and economically preferable. (1) Remove all subsidies from all energy supply sources—favorable tax treatment for oil, gas, coal, utility dividend reinvestments, investment tax credits, renewable energy tax credits, insurance subsidies for nuclear power, tax incentives for ethanol, and all of the others. (2) Deploy federal research funds so that conservation-efficiency research, renewable energy research, and coal-synfuels-nuclear research share equally. (3) Change all other rules so that buyers in each energy service market have a clear, informed, and unencumbered choice between conservation, more conventional energy, and renewable energy.

That listing is venturing into the area of energy policy, a subject treated explicitly in chapter 9. The issue is really more complicated than the preceding paragraph suggests, since short-run oil and gas prices are so far below their long-run marginal supply prices. With this qualification, the rules just given form the basis of energy policy proposals made later in this book.

Appendix   Nuclear Power in France

U.S. advocates of nuclear power correctly cite France as the Western nation with the most highly developed nuclear program. In other nations with rising outputs of nuclear electricity, most of the increase is from plants long planned and under construction and now being completed. Advertisements of the Committee on Energy Awareness (an American nuclear industry group) in 1985 cite fifty reactors ordered abroad since 1978 (there being none in the United States since that year). More than thirty of these are French; an examination of that nation's nuclear program should suffice for a review of the international scene.

The French nuclear program was adopted as the cornerstone of national energy policy after the 1973–74 oil crisis. Orders were placed at a peak rate of six reactors a year. Construction times were shorter than in the United States, standardized designs were adopted, several identical units were built at each site, and electricity costs appeared to be quite low. Nuclear power has risen steadily as a share of total electricity, and total national energy use, reaching two-thirds and a fifth, respectively, in 1985.

The French nuclear program involves much more than just building reactors. France mines uranium (though much is imported), refines it, enriches it,[31] and fabricates nuclear fuel units. Spent fuel is reprocessed to reclaim additional uranium and plutonium. A moderate-sized breeder reactor has operated for some years and a commercial, large-scale breeder, Super Phenix, is ready to operate.

Nuclear power has displaced oil in electricity generation, of course, but it also was intended to displace oil in other uses, especially space heat and industrial process heat.

As in all countries with energy plans or forecasts, future energy use in France has been revised downward periodically. Still, in the Ninth Plan (1983)[32] nuclear power was scheduled to provide 80 percent of France's electricity and 43 percent of its primary energy by the year 2000. This would be a highly electrified economy with electricity reaching far into markets long held by other fuels. France held to this course for about a decade. Now, in 1986, the future of the program is in doubt. The national utility, Electricité de France, has had large annual deficits for five of the past six years. Generating overcapacity is large and growing. Orders for new plants have been reduced to levels that the reactor construction firm, Framatome, finds unacceptable. Orders for breeder reactors beyond Super Phenix have been deferred indefinitely. Costs of nuclear plants are now rising as rapidly as they did in the United States, the United Kingdom, and West Germany.

In order to evaluate the significance of these developments it may be helpful to recapitulate the conditions for a viable nuclear program as the French designers of that program have long seen them.

One requirement for the successful functioning of a group of nuclear plants (risk considerations aside) is that they be embedded in an overall system that is functioning successfully in all areas. Uranium enrichment and fuel reprocessing must function economically. Permanent storage of highly radioactive wastes must be done routinely. There must be a healthy and economically viable industry for the construction, repair, and servicing of reactors. Wornout plants must be successfully and economically decommissioned.

All of these components must work, and their costs must be fully covered by the sales of nuclear electricity. The price of this electricity must, in turn, be fully competitive with all conservation-efficiency investments at the margin, and with all other energy sources, including renewable ones. France has the advantage of operating experience with fuel reprocessing and breeders as well as with fuel enrichment. Plant decommissioning and permanent waste disposal still have not been done, so that their costs are arbitrary estimates. Since most elements of the system operate successfully in France, the system would appear to be workable.

Yet, if that is the case, one may ask why is the growth of the system slowing down, why is the utility in such financial difficulties, and why is the reactor builder at the margin of economic viability?

Some of the answers are found in the recession that affects France along with the rest of the world, some are found in electricity markets, and others are found in the French system and its costs.

In industrial societies electricity is essential in about 8–10 percent of energy end uses, as Lovins, Taylor, and others long ago pointed out.[33] Given losses in generation and transmission of electricity, this implies an electric share of 20–25 percent in primary energy. About one-half of this is baseload electricity and therefore suitable for nuclear generation. With hydroelectric and pumped storage components in the grid, this share can be raised—perhaps to three-quarters, depending on the characteristics of a given system. If nuclear power is to provide more than 15–20 percent of a nation's primary energy, electricity must penetrate markets for industrial heat, space heat in buildings, or, with suitable batteries, vehicle fuel markets. This is well-known in France, where quite high nuclear shares are planned.

Yet, as chapter 2 pointed out, electric heat is very expensive. Oil can be priced as high as $100 per barrel and still be as cheap as electricity at 6¢ a kilowatt hour for industrial heat. Space heat costs can be reduced with heat pumps, but peaking demands on the utility are made much worse.

Industrial heat conservation is much cheaper than oil at $100 a barrel, not to mention competition from coal and natural gas. Conservation to reduce space heat demands in buildings is also cheaper.

These rather theoretical considerations may be illustrated by some examples from the recent French experience. Since the French nuclear capacity already exceeds baseload electricity demand (with ten more plants coming on line soon), coal is being displaced, not any longer oil. This distresses coal miners and adds to deficits in the state-owned coal mining organization. Nuclear plants attempt to go on or off more frequently to "follow load," but this reduces their capacity factors and raises the cost of their electricity.

In the space heat market Electricité de France is trying to sell electricity, while other agencies of the government are trying to deal with space heat in three other (competing) ways. Such conservation efforts as are officially made are trying to reduce space heat demands in all the ways mentioned in chapter 2. Such renewable energy efforts as are officially going on are promoting, among other things, district heating from warm geothermal resources that underlie nearly all of France. Still another arm of the government, concerned about the oversupplies of natural gas already contracted for, resists gas displacement, particularly in industry but also in the residential and commercial sectors.

France has seen a partial solution in exports of electricity. It must take care, however, not to sell electricity to other nations at prices lower than those charged to French industry, lest it further weaken the competitive position of French industry.

By 1983 the long-term energy planning group in the planning ministry concluded that no new generating plants would be needed well into the 1990s, even if rapid growth in electricity demand reappeared.[34] This would imply a long hiatus in new orders for Framatome, the French nuclear construction firm—for it, an impossible situation.

In a compromise, reactor orders were cut to two in 1983 and two in 1984. This was done mainly to avoid further unemployment in the reactor construction industry. Then, in 1984, the decision was made to order one reactor in 1985 and one in 1986. Reactors previously ordered and under construction would be completed at a slower pace. These measures produce some unemployment, a rate of orders that Framatome says is too low to enable it to maintain a viable entity, and an increase in costs per nuclear plant. The latter increase has been estimated at 20 to 40 percent.[35] Apart from this increase in reactor costs, real costs are already rising at a rate of 5 percent or more per year.[36] When outside observers adjust such data as are available to make them comparable with German, British, or American costs, the real cost increases turn out to be as high as in those countries.[37]

The nationalized utility, Electricité de France, lost $1 billion in 1982 and nearly that sum in 1981 and 1983. It owed $30 billion at the end of 1985, of which nearly half was in foreign currencies. Interest costs are burdensome. When the construction program is stretched, interest during construction for each reactor also will increase.

These factors and others suggest that French reactors ordered in 1984 will cost twice as much, in real terms, as those ordered a few years ago. That is, costs will reach levels that have created the present nuclear impasse in the United States.

Nuclear reactors have relatively high capital costs for reasons inherent in the technology. This is supposed to be more than made up by extremely low fuel costs. The French do not publish actual fuel costs, with which they by now have some considerable experience. They use only *estimated* fuel costs when showing the economic advantages of nuclear power. One large element in fuel costs is the charge for reprocessing. This, in the fuel cost estimates, is based on a reprocessing plant running as it "should" (80 percent of the time) and lasting twenty or more years. Since the reprocessing plant runs about 10 percent of the time, and wears out in six years, one can make some rough estimates of the experience-based actual cost for this component of fuel cost. France claims a much better operating experince in 1985. Time will tell whether this can be sustained. Also, fuel enrichment is highly electricity-intensive (16 percent of France's nuclear electricity goes to enrich uranium[38] at the European consortium plant, of which France is the chief customer). Since electricity is underpriced, especially at the margin, some adjustment needs to be made for this component of fuel cost as well. Operation and maintenance costs also greatly exceed estimates. Since the reactor model most used in France is similar to one of the most widely used types in the United States (Westinghouse), one would expect to find the same problems of steam tube corrosion, etc., that increase American operating costs.

When corrections are made for all of these components of the cost of French nuclear electricity, it turns out to be more expensive than coal-fired electricity,[39] and far more expensive than saved electricity from efficiency gains.

Electricité de France needs to raise its rates in real terms just to cover its existing costs and to amortize its huge debt. It needs to raise them still more to cover prospective costs. Newer reactors, for reasons already mentioned, will produce more expensive electricity than is now the case. Since they cannot run at full capacity, costs per kilowatt hour will be higher still. Yet the utility has been instructed to hold its price increases below the overall inflation rate. This was part of the compromise described earlier.

The utility and its vigorous construction program each year absorb 10 percent of all nonresidential fixed investment in France.[40] The marginal product of any further investment is zero, at least for a decade (perhaps negative when the displacement of coal and other factors are considered). Nations do not prosper by making large investments in zero-return industries.

Some of the choices that confront the highest level of French decision-makers follow: (1) Raise the price of electricity to reduce the drain of

EdeF deficits on the treasury. This will dramatize the high returns to conservation investments, disadvantage French industry, and undercut electricity exports. (2) Hold down the price of electricity to retain markets, and then face large and growing subsidies. (3) Stop new orders and cancel the most recently placed ones, thus allowing EdeF a chance to return to financial health and to slow the rise in its burdensome debt, but in the process destroying the reactor construction firm, 50 percent of which is owned directly by the state and 50 percent by the state-owned, deficit-ridden steel firm. (4) Build enough reactors to keep Framatome intact (four to six per year), creating ever-mounting excess capacity and ever-mounting debt for the utility and the treasury.

France has asserted from the outset that there were no halfway measures in the nuclear program—it was all (high energy growth, high electric, high nuclear) or nothing.[41] Now the government is trying to find a middle ground when, for a decade, France has insisted that there is none.

The best-informed observer of the French nuclear program in the United States, Jean-Claude Derian, had in 1978 outlined some of the potential problems in the French program, noting that it was proceeding with a "heavy political mortgage."[42] In a 1980 revision of his work he pointed out that the "first payments on that mortgage—if indeed, it even exists—have yet to fall due."[43] The program then seemed to be going well. As it turns out, he was less wrong than premature. The larger mortgage seems now to be the economic one, and that is now coming due.

These conclusions are not inconsistent with the U.S. experience. The cessation in orders came sooner in the United States, but institutions are quite different. The entire industry in France is government-owned except for a partial private interest in Framatome. In the United States, government owns the fuel-enrichment facilities (because of their sensitive nature with respect to nuclear weapons), while the rest of the system is mostly privately owned. Five U.S. nuclear utilities have come near bankruptcy; Electricité de France would be near bankruptcy too if it did not have access to the state treasury.

There remain two points of view as to who is "ahead" and who is "behind" in nuclear energy. From the point of view of the U.S. nuclear industry, France is far ahead. From another point of view, the United States is ahead, having first got in, and then, in an unacknowledged kind of way, been among the first to get out.

6

Other Industrial Nations

This chapter attempts to show the feasibility of an energy-efficient, renewable energy future for other OECD nations (those countries of noncommunist Europe, Japan, Canada, Australia, and New Zealand). The procedure is similar to that followed for the United States. Estimates of long-run efficiency gains in energy use are given and then renewable energy potentials are shown. The question of growth without energy constraints also is considered.

It is commonly asserted in the United States that Western Europe and Japan use energy much more efficiently than does the United States. The presumption often follows that potential efficiency gains are smaller than is the case in the United States or that no further gains are possible at all. As we shall see, matters are not quite so simple.

Most European nations and Japan do indeed use less energy per capita and per unit of gross national product than does the United States. This difference narrows when national products are converted at purchasing power equivalents rather than at exchange rates, but the difference remains. The definitive study in these matters is a Resources for the Future volume, *How Industrial Societies Use Energy*.[1] The United States and Canada emerge as relatively energy-inefficient. The United States uses more "energy services" per capita than do Western Europe and Japan and also uses energy less efficiently. (Using more "energy services" means, for example, driving more miles or heating more square feet of space.)

We shall first make some observations applicable to energy conserva-

tion-efficiency issues and renewable energy availability throughout the
OECD. We shall than examine, as case studies, Sweden, the United Kingdom, and Japan, offering more detail on renewable energy futures.
Every nation studied has shown remarkably large potentials for increasing energy productivity. National studies, varying in degrees of detail,
are available for West Germany, France, the United Kingdom, Denmark, Sweden, The Netherlands, Canada, and Japan. Sweden, for example, is already one of the most energy-efficient nations on earth. Yet a
two- to threefold reduction in energy use for its present level of output is
possible.[2] Tighter buildings, higher-mileage vehicles, and efficiency
gains in industry (especially steel and paper) produce this potential. The
West German study[3] takes efficiency gains even further, documenting a
three- to fivefold gain, but starting from 1973 levels of energy use.
Buildings there are not as well insulated as they already are in Sweden.
Savings in industry, in recycling materials, and in transport are pushed a
bit further. The United Kingdom study[4] outlines three- to fourfold gains
in energy efficiency as well—again through tighter buildings, more efficient appliances, higher vehicle mileages, and industrial gains.

In all cases, investments in increased energy efficiency produce higher
returns than investments in new energy supplies. These findings are so
consistent, and so universal, as to raise the presumption that they exist in
all developed societies. The Canadian studies[5] are in similar detail and,
of course, give similar results; I do not discuss them here because they
are so similar to U.S. results. The French and Japanese studies[6] are less
detailed, but yield similar results. Southern Europe (Spain, Portugal,
Italy, and Greece) are less well studied and use less energy per capita
than does Northern Europe. Levels of output and income are not as high
as in the north, but long-run energy use per capita should not exceed
efficient levels for Sweden, the United Kingdom, or Germany. Space
heat uses are smaller, and air-conditioning uses potentially larger, but
there are no other grounds for large differences, given similar levels of
income and industrialization.

These observations are supported by developments in energy use over
the period from 1973 to 1983 in the European Economic Community.
Output rose 22 percent, while energy use declined 8 percent. The energy-gross national product ratio fell at an average annual rate of 2.6 percent—a rate not far from the U.S. experience.

The outstanding case is that of Denmark. Over the decade energy use fell 17 percent, while output rose 20 percent. This was no accident; government policies with strong public support emphasized conservation-efficiency investments. The average annual decline in the energy-gross national product ratio, 3.75 percent, is one of the two highest in the world that has been sustained for a full decade; the other case is Japan.[7]

When we turn to Japan later in this chapter, we shall not look for a two- to threefold gain in energy efficiency for the long term. Opportunities abound, as the rapid efficiency gains in Japan since 1973 affirm. Still, Japanese energy use per unit of gross national product is already among the lowest in the group, in part because amenities in housing already taken for granted in the United States and Northern Europe have not become as widely available in Japan. Applying these energy-reduction coefficients to 1980 energy use in Japan (even though the Japanese study supports the possibility of large gains) would bias our results. There would be implied a very low level of "energy services"—a choice the Japanese may make, but one which outsiders should not assume.

OECD Renewable Energy Review

As to renewable energy, every OECD nation except The Netherlands has hydroelectricity. Most have the capability of some further expansion, even though the hydroelectric potential in Europe is more fully developed than anywhere else in the world. Official energy plans and forecasts typically assume that little or no more hydropower can be developed, yet capacity additions are under way in virtually all OECD countries.[8] National studies typically overlook gains from updating existing installations, and from small sites. The latter usually add 10–15 percent to conventionally estimated resources.[9] France already has installed some 400 megawatts at 1,000 micro-hydro sites. Suppose another 500 megawatts were available; this capacity would be negligible in high-energy, high-electric futures so frequently envisioned. In an energy-efficient France, 30,000 megawatts would suffice; one would not overlook 3 percent of that figure.[10] Switzerland provides an example of a nation increasing its hydroelectric capacity by upgrading old installations.

Tidal power, worldwide, is rather modest, but it is quite important in Canada, France, and (especially) the United Kingdom. Its awkwardness in timing (power is generated about six of every twelve and one-half hours, moving gradually around the clock with the tides) is mitigated by integration into grids with multiple generation sources and pumped storage.

The wind-electric potential is large in Scandinavia, the United Kingdom, coastal West Germany, The Netherlands, Japan, Canada, Australia, and Greece.[11] Geothermal resources are abundant in Japan, Iceland, Italy, Greece, New Zealand, and, for space heat, in France as well. When resource assessments are done, Spain and Portugal also may have geothermal potential.

Photovoltaic electricity is available everywhere but will first become economical in the areas with the largest annual number of sunny hours and multisource grids—Southern Europe, for example.

All nations have some biomass potential. Scandinavia and Canada are especially well endowed with wood. Even Japan, densely populated as it is, is two-thirds forested.[12] Crop wastes, food processing wastes, urban and animal wastes are available everywhere that people are.

Direct solar heat also is available in all nations, even though some have short winter days and overcast skies. Modern materials and techniques make possible collector surfaces that absorb heat readily but emit little of it. Such devices, technically available, are not needed for low-temperature space heat. That is more easily provided by passive solar design and superinsulation. Other low-temperature heat needs (residential and commercial hot water; industrial low-temperature heat) can be supplied by direct solar collection, with backup systems selected according to the relative abundance of other renewable sources. Geothermal heat systems may compete more effectively in much of Europe once the resource is more thoroughly assessed.

We shall shortly return to these matters with a more rigorous and quantitative analysis; this listing is merely to point out that renewable resources of all kinds are widely available and that even the least well-endowed regions have multiple resources. Complete self-sufficiency is not required. Trade in renewable energy sources is surely admissible where it is technically and economically feasible. Certainly coal-nuclear high-energy futures imply large volumes of trade in coal and uranium. The traded forms are most likely to be biomass-based liquid fuels and electricity—a pattern similar to that which now exists. Trade in goods no doubt will also reflect relative energy scarcities, as it does already in part. A showing of sufficiency in renewable energy for a given nation with reliance on trade, then, also must include a showing of at least one trading partner with a surplus of cheaply transportable energy sources. As an example, the United Kingdom will probably import biomass-based liquid fuels from France or Sweden, which nations have a much larger per capita resource base. The cross-channel electricity transmission line, already under construction, will simplify dealing with electricity demand peaks and matching up renewable electricity sources, though in neither nation is that an insuperable problem without links.

Illustrative Country Studies

Sweden.   In showing that a combination of energy-efficiency gains and development of renewable resources can meet long-run energy needs, we turn first to Sweden. We do this because the issues have been more thoroughly investigated there than in most countries so that better data are at hand. This version of an energy future has more official recognition in government and in official energy policy circles than in any other OECD nation save, perhaps, Denmark. National energy research budgets are therefore less skewed toward future technologies whose time has passed.

Sweden has a population of 8,300,000 with a total land area of 450,000 square kilometers (173,000 square miles). Land in crops and pastures totals 36,000 square kilometers; forests cover 250,000 square kilometers. Sweden, therefore, is less densely populated than most of Western Europe and Japan but has less cropland per capita. Its large forested area makes pulp, paper, and lumber its largest industry and provides a major renewable energy resource.

A detailed energy-efficiency study has been performed, with an abridged version published in 1983 in *Science*.[13] The introduction of energy-saving technologies in buildings, transportation, and in industry could reduce the 1975 energy–GNP ratio by 50-66 percent. Several growth scenarios are considered and corresponding energy demands are calculated with an input-output table and with the new energy-efficiency coefficients. In one, for example, output doubles over time, but housing and transportation grow less as a result of saturation. Some modest additional shifts to less materials-intensive, higher value-added products in industry also are assumed so that the remaining doubling of output occurs elsewhere in public and private demand. The result: total energy use declines more than 40 percent from the 1975 level, even with a doubled gross national product.

The method is, in some respects, similar to that used in the SERI study for the United States. Allowance is made for growth, there is some change in the composition (and energy intensity) of output, and total energy use declines (not only per unit of output, but in absolute terms). Sweden, however, starts from a lower level of energy use, and energy-efficiency gains are taken further than in the U.S. study. SERI asked, in effect, to what extent potential gains could be (partially) realized by the year 2000. The Swedish investigators, on the other hand, explicitly used all of the potential cost-effective and technically feasible efficiency gains. If the SERI study had done this, its conclusions would have been similar to my own, given in chapter 2. Sweden used, in 1975 and still in 1980, about 1.7 quads. This figure would be about 2.2 quads of primary energy, using U.S. measurement conventions (thus showing the importance of hydroelectric power in Sweden, and the large effect of stating electricity in terms of the fossil fuel that would have been used to produce it). Swedish practice states hydroelectricity at the energy value of the electricity only; conversion losses are recorded for fossil fuel-based electricity.

An energy-efficient Sweden with a doubled gross national product would use, on a basis comparable to the U.S., 1.4 quads. In another scenario explored more fully in the *Science* article, output would be 50 percent higher than it was in 1975, while energy-efficiency gains, not pushed as far as in the foregoing discussion, would permit energy use to fall about 35 percent from the 1975 (and 1980) level. This would come to about 1.5 quads as measured by U.S. practices.

The authors of the study then examine the renewable energy supply possibilities for this scenario. The point of mentioning both scenarios is that a renewable energy supply system that provides 1.5 quads would provide for either energy future. For the convenience of the reader who may wish to consult the original document, I will reproduce data from one of the tables (see table 9). The unit of energy measurement is the petajoule (1,055 petajoules = 1 quad).

The amount shown for hydroelectricity is nearly all in place already; the remainder is from official plans. The amount shown for wind is much below the potential. It might be developed only for export to a European grid (nearby neighbors Norway, Denmark, and Finland already have large potentials in hydro, wind, or in forest-products industry cogeneration). The biomass potential also is larger than the figure shown; 4 percent of the forest land, if devoted to rapidly growing species like

Table 9   Potential Renewable Energy Supply in Sweden.

Energy source	Energy supplied (petajoules)
Hydroelectricity	280
Wind electricity	40
Fuelwood and forest products; industrial wastes; agriculture	260
Other industrial wastes	5
Solar heat	5
All other biomass, including biomass farms	324
Subtotal—renewable energy	914
Metallurgical coal	36
Peat	50
Total	1000 (1)
Less conversion losses	100
End-use energy	900

(1) To convert to U.S. practice, add 560 petajoules for hydro "conversion losses"; this gives 1,560 or 1.5 quad primary energy.
Source: Adapted from Johansson et al., *"Sweden Beyond Oil."*

poplars and willows, could yield 350–540 pj. Sweden thus becomes a potential exporter of electricity and biomass-based liquid fuel, while allowing for an energy-efficient economy twice the size of the 1975 economy.

The United Kingdom.   We turn our attention now to a nation that seemingly would present insurmountable problems in constructing a renewable energy future—the United Kingdom. It has a relatively small area, a dense population, and a cool, cloudy climate. The population of 56 million inhabits 245,000 square kilometers—a density similar to that of West Germany, less than that of Belgium and Holland, but twice that of France and 3.5 times the population density of Sweden. Forests and woodlands amount to less than one-sixth of the area, cropland occupies about a quarter, and grazing lands take another quarter—statistics relevant to the biomass potential.

Since the United Kingdom, along with the United States and Canada, is not particularly energy-efficient, a large potential gain is to be expected. This seems to be especially plausible after examining the Swedish case. A recently published study by Olivier and others confirms these suspicions.[14] In their most striking scenario, gross national product triples, while energy use falls by one-half! This is not quite the sixfold efficiency gain that one might think at first glance. Some of the more energy-intensive components of national output grow less rapidly than the total, so that the changed mix of activities would not triple energy use without efficiency gains. When adjustments are made for these effects, a fourfold efficiency gain emerges. This is quite consistent with the findings for the United States discussed in chapter 2.

Britain now uses about 9 quads of primary energy. This amounts to half the energy per capita used in the United States. When per capita gross national product is measured at purchasing power equivalents and not exchange rates, Britain uses about 57 percent of the U.S. energy input per unit of gross national product.[15] The study just quoted implies that the present U.K. gross national product could have been produced with 2.25 quads if all of the energy efficiency measures already had been taken. Even in Britain that figure is so easily met that we will proceed to a more challenging level of renewable energy availability.

Following Olivier's scenario, we inquire into the possibilities of finding 4.5 quads in Britain, enough energy to support a gross national product three times as large as the present one. Since about one-third of this energy demand would be for electricity, and since the paramount concern of 1985 British energy policy seems to be building nuclear electricity plants, we first examine the renewable electricity prospect.

British hydroelectric capacity is not large, certainly as compared with most OECD nations. It has a hydroelectric potential of 2,500 megawatts, most of which is already developed. The small-hydro potential has been estimated at an additional 15–20 percent.[16] Some 1,681 megawatts of new pumped storage has just been completed. Several other sites with similar capacities are available but are not needed with the present configuration of Britain's electricity-generating capacity.[17]

Harnessing tidal power with a barrage across the Severn estuary has been discussed for years. This is hardly a trivial energy source; the 7,200-megawatt potential would furnish about one-fifth of the total electricity needs of an energy-efficient Britain. Since the capital costs per kwh compare favorably with those of nuclear plants and the fuel is free, British observers are somewhat puzzled at the lack of activity.[18]

The wind energy potential is very large—by some estimates as large as the present consumption of electricity.[19] The Electricity Board has had little interest, citing the variability of wind generation. This is a puzzling attitude, since utilities have long coped with many small and variable uses of electricity on the demand side. The hydro-pumped storage components would have to cope with variations in output from wind, tidal, and also photovoltaic sources—not an impossible assignment, but one made easier by the cross-channel tie with the French grid (France has abundant hydropower, as well as pumped storage sites and demand patterns that differ from those of the United Kingdom.).

Industrial and commercial cogeneration would provide 6,000 to 9,000 megawatts in the Olivier study—all in situations where the industrial technology of choice suggests cogeneration, not just a device for meeting electricity demands. The renewable electric potential, then, offers such a rich set of choices that 1.5 quads is easily found.[20] If, on the other hand, one were trying to find enough electricity to meet some official forecasts (nine times the Olivier energy-efficient level and twice today's capacity), it would be difficult to do with renewable sources. Such is the importance of taking the conservation-efficiency measures first.

Coal is used in the steel industry as in the U.S. and Swedish cases— perhaps one-fourth of a quad. Other renewable industrial heat would come from biomass or solar sources. Low-temperature heat for water comes from solar or geothermal sources or from biomass-based district heating systems.

Liquid fuels come mainly from domestic biomass sources, though strict self-sufficiency is not reached in the Olivier study. Imports of a small share (7–15 percent) are posited, as are "energy crops" from about 2 percent of the land area.[21] Overall biomass output comes to about 1.5 quads. British biomass sources include forests, agricultural wastes, animal wastes in dairies and feedlots, industrial wastes, municipal garbage, and sewage. The latter three sources vary more with population than with land area. U.S. estimates of the gross potential from these three sources come to about 6 quads.[22] On a relative population basis, this would imply about 1.5 quads for the United Kingdom. Lower per capita incomes would reduce the figure, though more of the gross resource would be collectible and usable in the United Kingdom. British agriculture and, especially, forestry are smaller than in the United States, but yields per acre in agriculture are higher. A British 1.5-quad potential for all biomass, including a small contribution from energy crops as assumed in the Olivier study, is quite credible. The remainder is solar,[23] additional electric (the renewable potential is so much larger than the strictly electric demand), or is imported, as dictated by relative prices.

As is the case in the United States, the realization of efficiency gains, even at a moderate pace, would so lower the demands on oil, gas, and coal that they would last much longer. New electricity capacity could be deferred, except for demonstration wind generators, and air pollution from existing coal plants could be greatly reduced simply by not running them. As is the case in the United States, official policies and energy budgets seem to be headed in precisely the wrong direction.

Japan.   Japan has experienced rapid gains in energy efficiency in the past decade and presumably has not yet exhausted that potential. We shall not, however, analyze Japan as we did the United States, Sweden, and the United Kingdom. We will establish a per capita level of energy consumption for a long-run, energy-efficient Japan that would entail amenities similar to those in the United States and Europe. It would be unfair to take present Japanese per capita consumption levels (already the lowest among OECD nations relative to gross national product) and lower it still further with regard to the efficiency gain potentials undoubtedly present. The Japanese, of course, may choose to do just that, but that is their choice. I would not want to bias my results by assuming Japanese amenity sacrifices.

If we take the U.K results just discussed and apply them to Japan, we get a long-run energy requirement of about 9 quads.[24] The population density (half again that of Britain but still less than The Netherlands) already reduces domestic passenger miles per capita in the transportation sector; this is one of the reasons why Japan consumes so much less energy for transport than does the United States or even Europe. Since the object of travel is to transact business or to provide personal pleasure, not just run up the count of passenger miles, we shall allow for some reduction below U.S. or European levels in the transport sector without necessarily implying a diminished standard of living.[25] We take 8 quads, therefore, as an estimate of energy requirements for a high-prosperity, energy-efficient future.

As now seems to be usual, the electric portion of this energy supply is readily found. Japan has a hydroelectric potential of 50,000 megawatts,[26] though not all of that can be developed. Present capacity is 20,000 megawatts with another 10,000 megawatts of pumped storage. Government projections already call for an additional 6,000 megawatts by 1990.[27] We take 1.3 quads (primary equivalent) as the long-run hydroelectric potential. The wind-electric potential is quite large, especially with offshore installations—so large that it will not need to be used except fractionally.[28] Geothermal resources are thought to exceed 20,000 megawatts.[29] When one considers the potential of a hydro-wind-photovoltaic grid alone, not to mention geothermal contributions and the biomass-based cogeneration that inevitably accompanies some industrial heat applications, renewable electricity potentials are abundant. Public transportation already is highly electrified, and will, no doubt, become more so. This will allow electricity further to substitute for liquid fuels, as it also must in a nuclear future. Electricity might provide as much as 4 quads, but a market penetration that high in an energy-efficient society seems unlikely.

The biomass potential has been estimated at 1–1.1 quads without energy farms.[30] As we have noted, two-thirds of Japan is forested, with the accompanying biomass potential centered on the forest products industry. Crop wastes, food processing wastes, animal wastes, garbage, and sewage provide the remainder.

Japan is half again as large as the United Kingdom but is much more densely populated. The agricultural sector is relatively small but very productive in yields per acre. More paper is recycled than in the United States or the United Kingdom, reducing the volume of municipal solid waste per person. Given the larger forested area than in the United Kingdom and the larger total area, the Japanese biomass potential may be understated.

Coal in steel production could amount to a half quad, depending on the amount of steel recycling and scrap.

Direct solar heat already is more widely used in Japan, relative to population, than in any country except Israel. Water heating in the residential and commercial sectors is sufficiently solarized so that Japan gets about 1 percent of its energy in this fashion. Some 2 quads of low- and medium-temperature heat from solar sources is not unreasonable in a long-run energy supply. Remaining energy demands, probably in the form of high-temperature industrial heat and liquid fuels, would be met by additional solar equipment, imports, substitutions of electricity, or by "energy farms"—depending on relative prices and the political tradeoff between degrees of energy self-sufficiency and food self-sufficiency.

Import Sources

The possibilities that Japan and the United Kingdom might import 5–10 percent of their energy requires that potential export capacity of transportable energy be shown. As an arbitrary choice, this is done without assuming Third World sources. We therefore look at the rest of the OECD group briefly. The United Kingdom could import (or more likely, exchange) electricity and liquid fuel, and Japan might import liquid fuel (electricity is not close enough). Only the most densely populated OECD nations (Japan, the United Kingdom, Germany, The Netherlands, and Belgium) are not likely to be self-sufficient in biomass-based liquid fuels. Imports to supplement domestic sources should not exceed some 1–2 quads. This fuel is readily obtained in Canada, Sweden, and Finland, all of which have large biomass potentials relative to their populations. It is implied, then, that the other OECD nations (France, Italy, Spain, and the United States being the largest) are at least self-sufficient. We have shown this to be the case for the United States in chapter 2. France, with energy demands similar to those of the United Kingdom (4.5 quads), would have about 2 quads of biomass by using various wastes and by devoting about one-half of 1 percent of its land area to energy crops.[31] Similar calculations show energy self-sufficiency, if need be, in Spain and Italy. Actual developments, no doubt, will be quite different. It suffices here to show the possibilities.

Europe and Japan, then, have attractive energy-efficient, renewable energy futures. With those futures come relief from heavy dependence on imported fuels that have loomed so large in energy policy discussions there. If one examines the price now paid for food self-sufficiency in the European Economic Community, it would appear that energy self-sufficiency could be bought less dearly, if that were thought to be important.[32]

7

Third World Energy Futures

Any study of renewable energy futures must reckon with the Third World. Three-quarters of the world's people live there, frequently in areas much more densely populated and less well endowed with resources of all kinds than is the case in industrial-market or centrally planned economies.

Even those official studies that have tardily recognized the possibilities of low or zero rates of energy growth in the wealthier nations still foresee enormous energy supply problems in the Third World.[1] This result would seem to follow almost automatically if one juxtaposes shrinking fossil fuel supplies with per capita energy usage moving toward European levels. Surely the "historic inevitability" of nuclear power is almost self-evident in the Third World, even if the wealthier nations could somehow live on renewable energy.

It is the purpose of this chapter to argue that renewable energy supplies are quite abundant and that Third World per capita usage at levels similar to an *energy-efficient* United States or Sweden can be supplied from renewable sources. This must be immediately qualified with respect to population—the above holds *provided* that population growth rates can be reduced and stable populations achieved by the time of another doubling. This qualification is not particularly damaging to our position, even given the magnitude of the problem. *Any* prosperous Third World future—with respect to food, housing, energy, or anything else—is equally dependent on population stabilization.

The reader will correctly perceive that my task is an ambitious one for a single chapter. The "Third World" is an abstraction, by no means homogeneous. Nations and regions differ enormously in culture, climate, population density, resource availability, income levels, and stages of industrialization, to mention only the more obvious characteristics. In one chapter I can at best persuade the reader that my conclusions are plausible and could withstand more detailed investigation. The chapter proceeds as follows: First, there is a sketch of the many forms of renewable energy available to nations seemingly poor in energy sources. This is mostly qualitative, with the aim of showing, for those not abreast of a recent but growing literature, that renewable energy is widely available. Second, some attention is paid to the potential for efficiency gains, though in general somewhat larger total energy inputs will be required to approach levels of prosperity now found in the OECD nations.[2] This is in contrast to our findings for the United States and Europe, where there was ample scope for economic growth with falling energy inputs. Next, I shall use as case studies the prospects for India and China, since nearly half of Third World peoples live in those two nations. Senegal, a small but representative African nation, also is examined. The last part of the chapter again looks at the Third World in the aggregate, with some rough calculations of energy inputs needed to match those of an energy-efficient OECD and the renewable sources to meet them. This last showing might seem unnecessary on grounds that, if India and China can achieve high income levels with renewable energy, everyone else can do the same. While this proposition sounds plausible, it is not necessarily so; things may be quite different in other areas.

A Brief Survey of Third World Renewable Energy Sources

A review of Third World renewable energy resources is instructive. The author has participated in numerous casual conversations on this subject, in which the following seems always to be assumed: Third World = tropical = sun = solar collectors. This is, of course, partially true; direct solar heat has a role in every nation. It also is very misleading. Third World nations cover the range from hot to cold, wet to dry, and cloudy to sunny. Other forms of renewable energy are sometimes more important than direct solar heat. Every nation has multiple sources; a few of the most fortunate have all of those listed below.

Hydroelectricity.   This already is the largest single electricity source in most Third World nations, yet the potential is much larger. Only 8 percent of the Third World potential has yet been developed, as contrasted with two-thirds in Europe.[3] Not every site will be developed; some are too costly or too remote or too destructive of fertile lands. On the other hand, the potential is probably understated, since some of the country estimates were made when oil prices were much lower, thus distorting the economic threshold for measuring the hydroelectric potential. In addition, small sites tend to be overlooked. In view of these possibilities it is not surprising to learn that a great worldwide boom in hydroelectric construction has been taking place. Between 1970 and 1981, world capacity nearly doubled, going from 291,000 megawatts to 484,000. About 123,000 megawatts were under construction in 1980, with another 240,000 planned.[4] In some respects the world has been talking nuclear but doing hydro. Brazil is a good example; the first of a planned ten reactors finally opened in 1982 and was shut down in 1984, having reached 50 percent of its design output only once. Meanwhile, Brazil's hydroelectric capacity grew more than fivefold between 1965 and 1981, to 31,200 megawatts. Its long-run potential, not fully appraised, is more than 90,000 megawatts, perhaps as much as 200,000.[5] Every Third World nation of any size has some hydroelectric capacity; among large populations only Bangladesh is modestly endowed.

Geothermal energy.   This, too, is widely available but in urgent need of a more comprehensive resource assessment. It is already known to be an important potential source in the Philippines, Indonesia, Thailand, Mexico, China, Turkey, the Arabian peninsula, Korea, Taiwan, Kenya, Tanzania, Peru, Chile, and in Central America. In the Philippines 1,500 megawatts of capacity is operating or under construction. That nation is second only to the United States in geothermal energy development.[6]

Wind energy.   Wind at velocities suitable for pumping water is available nearly everywhere. The stronger and more consistent winds suitable for generating electricity have not been thoroughly assessed. This is a grievous oversight, for detailed assessments can produce pleasant surprises. California was thought to have a very limited wind-electric potential until state-sponsored assessments in the 1970s indicated that a potential of at least 20,000 megawatts existed.[7] Wind assessments, along with similar resource appraisals for geothermal energy, would seem to be a high priority among governments and international institutions. The data already available suggest commercial-scale resources for wind electricity in China, the west coast of Latin America (Chile, Peru, and Ecuador), Morocco and other nations in the northwest corner of Africa, Egypt, and numerous islands—in the Caribbean and in the Mediterranean, for example.

Solar ponds.   Thermal-gradient saltwater solar ponds, first developed in Israel, can be used in most nations. They are most cheaply constructed and maintained near bodies of saltwater. Some occur naturally—they attracted little attention since they were not known, until recently, to be rich sources of potential energy. One natural site in Tunisia, for example, has an estimated capacity of 600 megawatts.[8] Since they are relatively land-intensive, they probably will be developed first in desert areas where alternative land uses do not have much value. Happily, they serve as complements to photovoltaic electricity in the absence of hydroelectric capacity. The Arab states stretching across North Africa from Morocco to Egypt and those in the Middle East, along with Israel, Pakistan, and Iran, seem best suited for initial development of this technology.

Tidal power.   Tidal power, as we have already indicated, is locally important, even though it is not a major resource worldwide. In the Third World, Korea, China, and India have sites that can be developed to advantage.

Solar electricity.   Photovoltaic electricity can be generated in virtually all inhabited places. The most productive applications are in areas with high insolation and proximity to hydroelectric or solar pond installations, i.e., most of the Third World. Other solar-electric technologies also may be competitive.

The reader will note that we have fallen into the custom of first considering electricity supply technologies, though electricity represents only a small fraction of energy end use. We thus seem to fall into the same trap as do nuclear-oriented energy planners who appear to think that any unit of energy is freely interchangeable with any other and that nuclear power can fill any "gaps" that appear. Actually, we are stressing the point that electricity can be provided from renewable sources in many ways. All of the above technologies, plus biomass cogeneration, plus other new technologies that may be developed, suggest that long-run electricity supply is a problem with numerous solutions. From this perspective, nuclear electricity is but one of many possibilities, and not a decisively attractive one at that.

Other energy forms: biomass resources.   The other forms in which energy is used, liquid fuels and heat, must come primarily from biomass sources and direct solar applications. Some transportation can be electrified (rails, urban trolleys, and buses) and some freight and passenger transport can be shifted to rails. For the remaining transportation demands, liquid fuels will be needed. Also, some heat can be provided through biomass combustion, especially in cogenerators, or from electricity, or from geothermal sources, but most heat will probably be provided by direct solar use.

Biomass resource assessment is under way but far from complete. A Resources for the Future study[9] estimates gross resources for Asia, Africa, and Latin America at 53 to 232 quads—the range indicating the degree of uncertainty in the estimates. Forest resources provide the larger share of biomass energy in most nations; the estimates given are based on existing forested areas. As deforestation proceeds, the estimates would decline. If reforestation can be accomplished and yields raised, the estimates would rise.

Biomass energy sources are available in some measure in all inhabited areas. At a minimum, they include human and animal wastes, the organic components of garbage, and some usable vegetation. In most areas they also include wood wastes, crop wastes, and crop processing wastes. In some areas energy crops, or multipurpose crops with an explicit energy component, are possible. These may include aquatic plants. Although wood dominates biomass resources in most nations, this may not continue to be the case as other biomass sources are developed.

It must be emphasized that we are discussing truly renewable energy. In biomass, therefore, we must take care to measure potential resources with due allowance for maintenance of soil fertility, erosion prevention, and the production of food, feed, and fiber.

Since wood is so important, some additional comments on Third World forests are in order. It is apparently the case in most countries that forests are disappearing at a rapid rate. Wood already is the major energy source in rural areas, but it usually is not being used on a sustainable basis.

Rural use is almost entirely for cooking, space heat, and conversion to charcoal. Wood is burned very inefficiently in all three uses. The adoption of more efficient cook stoves and more efficient charcoal conversion techniques would, in themselves, reduce wood withdrawals to a sustainable level in most of the Third World at present population levels.[10] The technologies are simple and available, but the sociology of their introduction is far from simple. This phenomenon should not surprise us; Third World villagers are remarkably like First World energy decisionmakers. Here the techniques of energy conservation and renewable energy are simple and available; the difficulties are in people's heads.

Successful reforestation is not a fanciful dream. The key seems to be involvement of and incentives to local people. Conspicuous success has been observed in South Korea and, on a more limited scale, in China and India.[11] Reforestation, like slowing of population growth, is a critically important issue worldwide. It is heartening to know that progress can be achieved in both areas.

Biogas production from animal and human wastes and from waste plant matter has several positive aspects. The energy content is partially removed in the highly useful form of methane, while nearly all of the nutrients and much organic material for soil conditioning are available, detoxified, to be returned to the soil.

Crop processing wastes and on-farm crop residues represent a considerable biomass resource—estimated at 14 quads in the Resources for the Future study mentioned earlier. Solid waste from urban areas will grow in importance if income levels outlined later should be attained.

Direct solar energy.   Direct solar heat is available everywhere, though insolation varies considerably from one place to another. Just as Florida receives less solar radiation than does Arizona, tropical Third World areas with some cloud cover receive less than arid areas farther north or south. Nevertheless, the average amount of insolation is high compared to Europe and Japan. Residential and commercial hot water, much industrial and agricultural process heat, and passive space heat can be provided in ways similar to those already discussed for the United States. In areas with heavy prospective air-conditioning loads, photovoltaic electricity is especially helpful, being well correlated with load. Solar units that use heat to drive the air-conditioning process also may be available.

The foregoing discussion has attempted to demonstrate that renewable energy resources are numerous, widespread, and large. In addition to solar heat, photovoltaic electricity, and biomass, available everywhere, nearly every Third World nation has at least two of the other renewable energy sources: wind, geothermal, tides, hydroelectricity, or solar ponds.

Later in this chapter we shall return to a more quantitative assessment of the resources in relation to long-run output levels. First, however, we shall look at China, India, and Senegal as important case studies.

Illustrative Country Studies

China. For these long-run purposes, let us suppose that the Chinese population levels off in accordance with current population policy in China and then declines back to one billion. We shall then depart from the approach used for the United States. We shall not, for example, ask what minimum energy inputs would have sustained the 1980 Chinese output given the introduction of all cost-effective efficiency devices and technologies (although that question is well worth asking—it could well yield an answer—down 75 percent—similar to the United States case).[12] We shall, rather, estimate a total energy input for a China as prosperous as Western Europe in 1980 in a method similar to that used for Japan in chapter 6. Our examinations of Sweden and the United Kingdom suggested that energy-efficient societies would use about 80,000,000 btu's per capita per year. In the United Kingdom case, this provided an allowance for further growth beyond a U.S.-Sweden standard of living. We adjust this downward to 50,000,000 btu's per capita in the Chinese case,[13] a bit more than we did for Japan, for the following reasons: (1) A Western European 1980 per capita output would be regarded as an incredible development success in China, and our larger per capita consumption figure was for Sweden and a richer United Kingdom, not the 1980 European average. (2) A wealthy but densely populated China would probably use less energy per capita for transporting goods and people—the same consideration that was applied to Japan.[14] (3) The share of heavy, energy-intensive industry in the 1980 Western Europe figures overstates long-run requirements—a phenomenon most visible in the steel industry.

We shall then skip over a treatment of present energy end uses in China and the scope for efficiency gains that would be followed by a large provision for growth to OECD income levels. We shall, rather, take one billion people with a per capita use of 50,000,000 btu's per year and ask whether China could produce, sustainably, 50 quads of renewable energy. Then it will be necessary to assess the "match" of energy sources and end uses.

A billion people crowded into a portion of the Chinese land area enjoying levels of prosperity similar to those of Western Europe in 1980 and all on renewable energy is not among the world's most plausible propositions. Yet the 50 quads of renewable energy estimated to be required for this outcome is readily found.

China has enormous hydroelectric potential. Estimates of outputs that are technically and economically feasible range from 13.5–22.5 quads (primary energy equivalent)—some 330,000–550,000 megawatts.[15] Since photovoltaic and wind electricity also are available to be integrated into a hydro-dominated grid (without, therefore, separate storage), and since geothermal, some tidal, and biomass-cogenerated electricity also are available, virtually the entire 50 quads could be produced as electricity if there were any reason for producing that much electricity.

For liquid fuels, and for some heat or electricity, large biomass resources are available. Gross resources have been estimated in the Resources for the Future study at 9–15 quads.[16] This is based on a forested area smaller than that shown in other sources. Further reforestation, which must be undertaken anyway if income levels are to continue to rise, would expand this estimate. More forest resources are needed for lumber and paper, to stabilize soils, and to avoid rapid siltation of hydroelectric reservoirs. Good management of forests, which would move yields to the high end of the estimates given, also is essential. I therefore take 20–22 quads as a biomass energy estimate—the higher Resources for the Future estimate, adjusted upward for additional forested areas, some energy crops, and adding in urban wastes.[17]

Liquid fuels would represent a smaller fraction of end-use energy than in the United States, since the downward adjustments in per capita energy use from U.S. levels are mainly in auto transport. Electricity and heat therefore would take somewhat larger shares. Industrial energy would be a larger share of the total, as is already the case in Japan.

Rough estimates, building up from end uses in residences, commerce, industry, and transport, come to 20 quads of electricity, 20 quads of heat, and 10 quads of liquid fuels. Sixteen quads of the gross biomass potential would be required for liquid fuels, allowing for conversion losses and process energy.

The electricity demand is more than adequately provided for; relative costs would determine the balance between hydroelectricity, geothermal electricity, biomass cogeneration, photovoltaics, and electricity from wind and tides.

In the heat budget, residential and commercial space heating and hot water would take 6–7 quads—mostly from direct solar sources. Available backup sources are low-temperature geothermal heat, biomass fuels, or electricity.

In the industrial sector, steel production would use metallurgical coal for 2–3 quads. This leaves 10–12 quads of heat demand to be met—mainly industrial, but also cooking in the residential and commercial sectors. Available sources are the remaining 4–6 quads of biomass resources, direct solar heat, and, in specialized cases, electricity. It is not necessary to specify precisely the shares of low-, medium-, and high-temperature heat, since the solar share would not have to be more than half. The rest is biomass or electricity.

This quick review cannot do full justice to an issue as comprehensive and complicated as that of energy supply for the world's largest nation. It does sketch the main features that would be covered in any more extensive study and gives an initial demonstration of a feasible energy-efficient, renewable energy future.

India.   In some respects China is an easier case than India, partly because it has made rapid strides in reducing its rate of population growth. It also has larger hydroelectric and geothermal resources per capita. Nevertheless, India, especially an India with population growth falling as rapidly as in China, has interesting possibilities.

I take, as a basis for long-run calculations, a population of 800,000,000. Clearly, India's population will surpass that figure unless there is a drastic slowdown in growth rates. Since no energy or economic or any other kind of future makes sense without a population somewhere in that range, my figure may be classified as wishful thinking (in a class with Indian nuclear planners). It also may be regarded as a goal to be passed and then approached from above, as is the case in China.

This may not, perhaps cannot, happen. Yet if it does not, there is no reason even to consider the energy needs of a high-income India, for there will not be one. Whatever may happen, India will be no worse off using energy efficiently and using renewable energy than doing anything else.

All of the above implies an energy supply of some 40 quads, to use the per capita figure already developed for China. As usual, we start with electricity.

India has 70,000–100,000 megawatts of hydroelectric potential,[18] with an annual output equivalent to some 3–4 quads primary equivalent.

Wind resources have not been assessed very thoroughly, but are not promising in areas where they have been measured, at least for electricity generation (water pumping, though, is a nontrivial use of windpower). Geothermal explorations are just getting under way, with areas of most promise in the northwest. Tidal power is available at several sites, with a potential of 4,500 megawatts.[19] We have not in this study relied on OTEC (the use of ocean temperature differentials to generate electricity) since the pilot scale devices have not operated for as long a time as the technologies used here. Nevertheless, should this become a viable electricity source, India is one of the nations with the requisite temperature differentials in the Indian Ocean off its southern coast. Solar ponds are available in some locations. Nepal, with some 80,000 megawatts[20] of hydroelectric potential, may become an exporter of electricity to India and Pakistan, as well as a future home of part of the world's aluminum industry.

An electrical system built around hydroelectricity, photovoltaics, geothermal plants, whatever windpower may be found, biomass cogeneration, and with some contributions from tides, solar ponds, and possibly OTEC, should be capable of supplying 10–12 quads (primary equivalent) of electricity, especially if imports from Nepal are considered.

The biomass potential is estimated in the Resources for the Future study at 8–15 quads. The higher figure takes the forested area (which, unfortunately is shrinking) and applies standard yields. The lower figure relies very little on forest outputs and is based mostly on crop residues and animal wastes. Since reforestation is essential, as in the Chinese case, to a high-income economy for which we are estimating energy sources, we take 3–4 quads in addition to the higher figure above. The forest cover is needed to avoid soil erosion and reservoir siltation and to provide wood products as well. Fuel for cooking, a major present use of rural area forests, is potentially available from biogas installations. We can add another 2 quads for aquatic plants and for urban wastes, not included in the above estimates. A biomass figure of 20 or 21 quads is thus attainable.

India's climate permits the use of aquatic plants in wastewater purification—a technology that increases the energy potential of urban wastewater and leaves residues, detoxified, for fertilizer use.[21] Energy crops also may be profitable, especially "energy cane"—India already raises sugarcane.

The biomass resources, as in China, must provide liquid fuels for transport as well as gas for cooking and for high-temperature industrial heat. The liquid fuel component is thus 8 quads net, or about 13 quads of the gross biomass resource allowing for conversion losses and process energy. The remainder of the biomass resource, 7–8 quads, is available for other uses.

If electricity of 10–12 quads, and biomass of 20–21 quads are provided, an additional 8–10 quads of direct solar heat would complete the primary energy total of 40 quads estimated to be demanded. Steel manufacture would get about 2 quads of heat from coal, thus reducing the solar heat requirement or adding flexibility to total supply.

Space heat demands are smaller in India than in China and would be met largely by passive design. Three quads of heat for residential and commercial hot water lends itself to solar supply, mostly from rooftop collectors. An additional 3–5 quads of solar industrial process heat would be provided. Other industrial heat demands and cooking demands would have 7–8 quads of biomass available, since liquid fuels would not take the entire biomass supply.

Electricity supply is less readily met from hydroelectric sources than in China. The biomass resource per capita is somewhat larger, suggesting a greater reliance on biomass-based cogeneration. More photovoltaic electricity might be used, along with solar ponds that provide some of the required energy storage. Pumped storage also might be required in addition to the hydroelectric component of the system.

Developing the full 40 quads of renewable energy will not be easy in India. But then, it would be much easier than providing the 120 quads or so of conventional energy implied by present development plans.

Other Third World nations.   The reader is entitled to expect other evidence for viable Third World energy futures—perhaps some examples from Latin America and Africa.

Latin America is so richly endowed with renewable energy sources that no demonstration is made here. In addition to the ubiquitous biomass and direct solar sources, Latin America has high per capita endowments of hydroelectric, geothermal, and windpower resources. Its challenge is population growth, not available energy.

Africa may be a different matter. For that reason, an African example is given.

Senegal.   Senegal is a small country, but it has many characteristics of other African nations. Part of the land is semiarid, forest cover is diminishing as fuel wood is taken, population is growing rapidly, and 80 percent of the work force is in agriculture.

Overgrazing and deforestation combine to encourage the advance of the desert—a situation found in the belt of nations across Africa just below the Sahara. Another part of the country is more like tropical Africa, having more abundant rainfall with tropical forests and tropical agriculture.

The approach to the consideration of an energy-efficient, renewable energy future is the same as for India and China. A stable population figure is selected, a per capita energy use for a high income society is applied, and a total energy demand is thus estimated. Then the various forms of renewable energy are reviewed, their potential quantities assessed, and a matching of sources to end uses is analyzed.

Senegal's 1983 population was estimated at 6.2 million. The growth rate is high and shows no signs of slowing. The World Bank, using some assumptions about eventual slowing, gives a hypothetical stationary population of 30–36 million some time in the twenty-second century.[22] The year 2000 projection is 10 million.

Given present resource endowments and present technologies, there is no point in discussing a high-income Senegal of 30–36 million people, for there would not be one even if most of the capital involved were free. An arbitrary choice is 15 million; this implies a slowing of population more characteristic of the Chinese experience than of other developing nations.

The next step is to estimate total energy demand for a high-income society. The Japanese-Chinese figure of 50 million btu's per capita per year is used. This implies a primary energy demand of .75 quad, or, in more manageable units, 750 trillion btu's. Could that be provided from renewable sources in Senegal?

The electricity potential is first examined. The electricity component of demand in energy-efficient societies suggests a demand of 250–300 trillion btu's. World Energy Conference figures show a hydroelectric potential of 4,400 megawatts, or about 200 trillion btu's. The World Bank gives a much more conservative estimate of the developable potential—500 megawatts or 25 trillion btu's.[23] None of the hydroelectric resources are now developed.

Senegal is one of the African nations with a wind-electric potential. It also is well suited for the use of solar ponds—a useful supplement to hydroelectricity for baseload or peak electricity. Photovoltaic electricity and biomass cogeneration complete the listing of renewable electricity sources.

The hydroelectric potential is taken to be 2,000 megawatts, or 100 trillion btu's. The smaller World Bank figure seems to be excessively conservative. Some 50–100 trillion btu's of wind-electric and photovoltaic electric capacity could then be used in conjunction with hydro and other storage capacity.[24] The remaining 100 trillion btu's of electricity would be provided by solar ponds or biomass cogeneration, the latter as determined by the energy demands of the industrial sector and the heat-electricity cogeneration potential.

For estimation purposes, and to provide consistency with biomass supplies and other demands, we assume a 50/50 division of the remaining electricity demand. This gives 1,000 megawatts each of cogeneration and solar pond electricity.[25] The latter complements the hydroelectric facilities in storage.

Biomass sources would need to provide for liquid fuels in transportation, heat (mainly industrial) not provided by direct solar sources, and the electricity cogeneration demand added to the industrial heat demand.

The appraisal of the biomass potential begins with forests. About one-third of the land area was forested in 1980. Most of the removals are for fuel wood, which is burned inefficiently in primitive stoves or is converted (again very inefficiently) to charcoal for urban cooking. Removal at present rates would destroy the remaining forest cover in about thirty years.[26]

This forest area under good management would yield, sustainably, half a quad or 500 trillion btu's. Allowance for lumber and pulpwood demanded in a high-income society would reduce the portion for energy to perhaps 250 trillion btu's. More productive use of existing grazing lands, along with reforestation, would offset the larger crop- and grazing-land requirements of a larger population. We use 150–200 trillion btu's as the forest biomass energy estimate.

Other biomass sources are crop wastes, animal wastes, and urban wastes. Total biomass sources come to 300–380 trillion btu's without energy crops.[27]

The remaining energy demand of approximately 100–200 trillion btu's is provided by direct solar heat. This is consistent with both the remaining gross energy demand and end-use patterns. Hot water, space heat where needed, and much industrial process heat lend themselves to solar supply.

It remains to be shown that biomass sources can meet the probable liquid fuel demand and still provide for cooking, high-temperature industrial, heat and cogeneration of electricity.

With urban transport and intercity rail electrified, auto and other vehicular transport would demand about 150 trillion btu's, or 250 trillion gross, before conversion losses and process energy. The total biomass resource has been estimated at 300–380 trillion btu's. Some energy crops may become necessary to provide for liquid fuels and to meet other biomass demands. Just 1 percent of the nation's land area planted in energy crops would yield 50–100 trillion btu's. This flexibility would seem to accommodate any insufficiency of biomass from waste streams and forests.

In the Senegalese climate, as in most of sub-Saharan Africa, the use of aquatic plants to purify municipal and food-processing waste water offers an integrated water-recycling, energy-producing, and nutrient-recycling system. Aquatic plants like water hyacinths have some of the highest per acre yields on earth; energy production per capita from waste water can be doubled.

In the case of Senegal, once again, there are numerous ways to provide renewable energy. The particular set chosen would depend on related decisions in forestry, agriculture, and waste treatment. It suffices to show that at least one set of sources meets our estimate of energy demand.

These three case studies can be generalized to a showing of world renewable energy sufficiency. That step is taken in chapter 8.

Still, it is only a kind of "existence theorem." This is an important step—one cannot talk about paths forward toward a given energy future unless there is some reason to think that it exists. Then one can begin to think about actions and policy measures to encourage economic development with an energy-efficient, renewable energy component.

An interesting perspective on such an energy future is given in the Princeton study cited above.[28] The authors argue that this kind of energy development is consistent with a development program that aims to meet basic human needs, develop rural areas, provide productive employment and greater equity in the distribution of income. It has the enormous extra benefit of avoiding nuclear proliferation and the buildup of carbon dioxide in the atmosphere.

Chapter 9, which deals with energy policy matters, includes a section on the policy implications of our analysis for developing nations.

8

Some Notes on the Transition

This chapter first extends the results of earlier chapters to the entire world. Energy demand and renewable energy supplies are examined for an energy-efficient world with a stable population of eight billion. A growth path of output and of energy that might lead toward renewable energy sufficiency is examined for feasibility. Economic growth can be accommodated without recourse to additional coal and nuclear power if the world chooses to do so. Energy efficiency gains at rates already achieved, and deployment of renewable energy at rates also already achieved would be required—developments already grounded in experience.

Such a growth path, developed only to show that one such path, at least, exists, requires energy-efficiency gains and renewable energy development at rates experienced mostly with historically high oil prices. In an era of falling oil and gas prices, conditions may not favor this kind of energy future. The outlook for energy prices and their probable ranges are next examined. This sets the stage for a discussion of energy policies in chapter 9.

A Renewable Energy World?

Chapters 2, 3, 6, and 7 have shown that the United States, some representative industrialized countries, and some selected developing nations could support prosperous economies with renewable energy sources, providing that populations are stabilized. Can this proposition be shown to hold for the entire world?

The answer can be approached in the following way: Establish a population target at which world population levels off—say eight billion.[1]

In a relatively high-income and energy-efficient world, per capita energy use would be somewhere between that of the United States (chapter 2) and China (chapter 7) (the latter at income levels now found in Japan and Western Europe). This implies a world demand for energy almost twice the present use—480 quads.[2] This is, of course, a much lower level of energy use than is generally forecasted, since so much efficiency gain has been entered into the calculation.

Can 480 quads be supplied from renewable sources? Clearly so. Investigators for the International Institute for Applied Systems Analysis,[3] in a world energy study, identified 1,000 quads of renewable energy supply. Their estimate relies relatively little on direct solar energy, especially solar electric contributions. Other investigations have been equally optimistic,[4] but I rely on the IIASA estimates, since the main point of that study was that future energy demands would be very large and that coal and nuclear power would be the main suppliers. That is, it was not a study biased in any way toward renewable energy.

The world figures, then, confirm the conclusions of the country studies. Some regions or nations would import fuels and electricity, and others would be exporters. Overall, supplies would be adequate.

All of this, so far, is again a kind of feasibility exploration or "existence theorem." This is a very important step. If one does not know that a particular state of the world is possible, or believes it to be impossible, he will not consider policies to attain that state, even if it might be thought desirable.

A Numerical Sketch of the Transition

The next exploratory step is to see whether one can get there from here. Problems are approached throughout this book by subdividing them. That approach is continued here. First, a simple dynamic model is used. It shows that, under plausible assumptions grounded in experience, the world can indeed move steadily toward an energy-efficient, renewable energy future. Moreover, and this is a major point, it can do so without increased reliance on coal or nuclear power. These are widely regarded as the fuels of the future, or at least the bridge to the future. They may be used, but they need not be; there is a choice. The world economy, from an energy perspective, can grow at rates like those of the 1960s and can meet its energy demands over time with oil, gas, and gradually increasing outputs of renewable energy. This time path recognizes the eventual decline in the outputs of oil and gas. There are many reasons why economic growth rates of the fifties and sixties will not be easily reattained. Energy constraints need not be among these reasons.

World Growth Assumptions

The World Bank estimated that a "high growth" path for the world, 1985–1995, would be about 4.6 percent annually.[5] With population growth nearing zero in the industrial nations, long-term economic growth rates there are likely to decline after 1995, to the 2–3 percent range.[6] Developing nations could thus grow very rapidly with a world average rate of 4–4.5 percent after 1995. As an example, if developing nations in the nineties have a third of the world's output, they could grow at 8 percent and still leave a world average of 4 percent, if the other two-thirds of the world economy were growing at 2 percent.

As population growth rates in developing nations decline (and they must if high incomes are to be attained), then slightly lower world average growth rates (in the 3 to 4 percent range) would still permit rapid "catching up" in per capita incomes in the developing world.[7]

Could the world grow at 4.5 percent for a decade and 4 percent thereafter while moving steadily toward more efficient energy use and reliance on renewable energy? A numerical exercise can give a simple but adequate answer to this question. The same analysis can then be modified to fit more complex situations.

Improvements in energy efficiency have been maintained at rates of 3 to 4 percent for periods as long as a decade in some nations. Suppose that rates of 3–3.5 percent could be maintained worldwide.[8] This would permit economic growth at rates of 4–4.5 percent with energy inputs growing at .5–1.5 percent. Economic growth at historically high rates can thus be attained with little growth in energy inputs. Add to this an expansion of renewable energy inputs at a rate equal to one-half to 1 percent of world energy use—a very modest rate.[9] Then economic growth can proceed, fueled mainly by gains in energy efficiency and by increases in renewable energy. Any residual energy demands could be met for some decades by a modest expansion of oil and gas use. Since world oil output in 1985 was about 20 percent below its capacity, the kind of growth pattern just described does not strain world oil markets.

Oil and gas output will, at some point, reach a peak and then decline. That point already has been reached in the United States, and for oil in the North Sea it probably will be reached in the late 1980s.[10] Oil output in the Soviet Union also may have peaked. These observations have implications for the nations involved as they consider their energy policies. The world peak may now be deferred until after the turn of the century,[11] contrary to expectations in the 1970s. Had petroleum use continued to grow at rates seen in the 1960s and early 1970s, the peak would have come sooner. It is the decline in petroleum use that has created excess capacity in the industry and offers the possibility of several decades of modest worldwide output expansion.

The prospect of declining output of oil in the United States leads, in official projections, to a rise in oil imports in the 1990s. This is a direct result of assumptions concerning continued economic growth and a slackening in the rate of gain in energy efficiency.[12] That assumption is, in part, grounded in the falling price of oil. These matters, important for U.S. policy, are pursued in chapter 9.

It suffices here to point out that world growth can continue without energy constraints at rates near historic highs. Rates of efficiency gain that already have been sustained in several countries are required, as is the continued deployment of renewable energy sources. In the first few decades some increase results in dependence on oil and gas, but this increase need be no more than is readily accommodated by world excess capacity. After the oil peak, an acceleration in the rate of renewable energy deployment is required.

Evidence for 3 percent Annual Efficiency Gains

We now need to review the evidence that efficiency gain rates can be maintained and that renewable energy sources can be put in place rapidly enough. The most rapid sustained efficiency gains have taken place at rates of 3 percent and 3.75 percent, respectively, in Japan and Denmark.[13] In these two nations energy conservation-efficiency has been an integral part of energy policy since 1973. Japan has made a conscious move away from energy-intensive industries in which it had no long-run advantage and has systematically implemented energy-saving investments in industry.[14] Annual average rates of this magnitude, sustained for a decade, produce dramatic drops in the energy–gross national product ratio—25 percent and 30 percent, respectively. The United States, incidentally, has managed to drop its energy–gross national product ratio almost as fast—25 percent in twelve years, and for the 1979–84 period, has matched the Japanese and Danish experience. It can continue to do so if the policy measures outlined in chapter 9 are adopted.

What has already been done, then, cannot be dismissed as impossible. Maintaining this pace of "energy productivity" gain, incidentally, doubles a nation's energy efficiency in two decades and quadruples it in four decades. In principle, still more rapid gains are achievable in the centrally planned economies, since energy use relative to output is two to four times that of the Western industrial nations.[15] Even though developing nations use much less energy than the industrialized nations, they use it relatively inefficiently. There is ample room for efficiency gains there, too. Some of the most productive investments available are those that reduce energy requirements for existing outputs.[16]

Evidence for Renewable Energy Growth

Our simple model also involved increasing reliance on renewable energy—in fact, adding 1–3 quads per year. These are modest numbers indeed if one looks back at energy projections of the 1970s and the very large additions to energy supply that they contemplated. Can these figures be related to gains actually experienced? If one looks at rates of addition to world hydroelectric capacity in the last decade, and looks at rates of development for other renewable resources in the most successful nations, the answer is clearly yes.[17]

A Conclusion for Coal and Nuclear Power

The conditions laid out for the simple economic-energy growth model can be met—a conclusion based on observed achievement, not estimation or speculation. The conclusion is so important that it bears repeating: *the coal-nuclear "bridge" to the future may be bypassed altogether, and the world is free to move, in an orderly manner, slowly but directly into a renewable energy future.*

This is one path. It may not be chosen. It is important to know that it exists. In fact, actual developments probably will be quite different from anyone's model—simple or complicated. My model captures the most important elements of any such model and it is related to the experience of the past decade. If the preceding exercise had been attempted, but could not be shown to work, then the choice of possible energy futures would be narrower by one.

Trends and Cycles

In many important respects the world does not move in the way models usually specify. In particular, it moves in fits and starts—in cycles around long-term trends, whereas models usually assume steady movement along long-term trend lines. Nearly all energy-economic models assume a steadily increasing price of oil, which price increase drives the other variables. That, indeed, is implicit in the above analysis for efficiency gains and for renewable energy investments.

In 1985 the price of oil fell drastically, and it remains in 1986 well below the long-run marginal supply price given in chapter 2. Since 1973 the real price of oil has twice jumped, and then sharply declined. For nations affected by the appreciation of the dollar, there has been an additional period of rising, then falling, price. A new phenomenon has appeared—the world oil-price cycle.[18]

The Iranian revolution and the following invasion of Iran by Iraq produced the second jump, sending oil prices to new highs. The OPEC price per barrel went from $13 in 1978 to $34 in 1981, and spot prices went over $40. From several perspectives[19] these prices were unsustainably high, though only a few oil market analysts were sufficiently confident of their analysis of the situation to say so publicly.[20] The first confirmation of these views came with a developing oil surplus in 1982, and a drop in the OPEC price in 1983. Prices have continued to erode since then, and the "temporary" surplus continues. In mid-1985 world oil demand was still running at rates well below 1979 demand (53 million barrels a day vs. 62.5 million).[21] Since non-OPEC production has risen by about 5 million barrels a day over the 1979–85 period, OPEC (disproportionately Saudi Arabia) has reduced its production by about 15 million barrels a day.

Oil price fluctuations are a mixed blessing. During periods of declining prices, inflation rates are reduced and growth rates increased. Upswings in oil prices are inflationary and output-decreasing. Given the institutions of our economy, these effects are not symmetrical but tend to produce permanent increases in the price level, and, perhaps, raise the level of unemployment consistent with a constant inflation rate. In the present phase, with oil prices again at a low point, there is an additional political effect: the public believes that any energy problems that may have existed have been solved. The result, for our purposes here, is that falling oil prices retard the rate of energy-efficiency gains and slow the pace of renewable energy investments. They also retard the development of synfuels, coal use, and nuclear power.

Cycles around long-run trends are not unfamiliar phenomena. They long have existed for total output, prices, and individual product outputs. The nature of our problem is to discover an appropriate time path for a long-term energy adjustment process, to investigate the implications of cycles in oil prices for that process, and to see what implications for energy policy emerge.

Oil Prices: Long-Run and Short-Run

In order to assess the impact of these variables on our analysis, we need to distinguish between the long-run marginal supply price for oil (which determines its long-run price trend) and the short-run marginal cost of producing oil (which sets a lower limit on short-run prices). With this distinction in mind, we can ask whether the long-run price used in chapter 2 has changed, since that would change all of the estimates of the conservation potential. Then we need to estimate the probable lower limit of short-run prices, which can be well below long-run prices for some time.

The short-run price of oil must cover at least the cost of extracting it from existing wells—that is what is meant by the short-run marginal cost. If all oil came from Saudi Arabia, that cost might be quite low—$1 or $2 a barrel. But even in a depressed oil market, 53 million barrels a day are being sold. It is the cost of getting the fifty-third million, not the first million, that governs the price. Extraction costs and other operating costs vary considerably from one area to another. The costs are rather closely guarded information in the oil industry. Estimates for the North Sea wells run from $4.00 to $8.00 per barrel, and for Alaskan wells,

$10.00 to $15.00.[22] Many U.S. oil fields, especially those using enhanced recovery techniques, would not operate in a truly competitive world oil market. Their output would be replaced by cheap Arabian oil. Estimates of a working figure for short-run marginal cost at an output of 53 million barrels are about $10.[23] Prices could go below that figure for a short time, since something must force marginal producers to shut down. Prices would not remain below $10 for very long; no one will produce oil if out-of-pocket operating costs are not covered.

Even though prices in mid-1986 were fluctuating in the $9 to $15 range, there is reason to believe that they will return to the $20 range. There may be a period of jockeying for position by the various suppliers before this price increase takes hold. The reasons for this assertion about a $20 price will be given in a moment.

The long-run marginal supply price for an output of 53 million barrels per day is estimated in a different way. One must ask what sum would be required per barrel, at the margin, to locate and make ready for production new wells capable of an output of 53 million barrels per day. This is quite a different matter from operating existing wells. One can readily see why the long-run price is higher than the short-run price under excess capacity conditions. Moreover, the real long-run price is likely to rise through time as successively more remote and difficult terrain is explored and as the quantity of remaining oil diminishes. This tendency might, of course, be arrested by the availability of good oil substitutes—fossil synfuels, for example.[24] As chapters 2 and 5 have pointed out, these prices are well above the long-run oil price used in chapter 2 for 1980.

If OPEC were able to operate successfully as a monopolist, then there would be a price path that would maximize net revenue over a long period. A calculation of that price path has been made, and suggests a price of $26 to $32 in 1985, gradually rising to $55 in 2000.[25] That price would then become the operative short-run price. Since OPEC apparently is unable to operate as a monopolist, such a price is not relevant. Still, for Saudi Arabia alone, in a more limited sphere, there is a revenue-maximizing price. By estimating world demand through time, the supply responses of non-OPEC producers and the supply responses of other OPEC members, Saudi Arabia can control its own output in a way to maximize its own revenues. That takes place at a lower price than an OPEC-wide output and price decision. The relevant price has been estimated at $20 per barrel in 1985.

My own guess is that the price will fluctuate for another six to twelve months as Saudi Arabia expands its sales and forces the shutdown of high-cost wells. These, incidentally, are mostly in the United States. Then we are likely to see the price move toward the $20 figure. This is the view of S. Fred Singer who long ago correctly predicted the decline in oil prices. These developments are also consistent with the views of Morris Adelman if his "statesman's shadow price" is used.[26]

If the short-run price is expected to stabilize soon in the $20 range, is there any reason to change our earlier estimate of $34 (1980 dollars—$40 in 1985 dollars) used in chapter 2 for the long-run price? Only if falling oil prices somehow lower the price of locating new oil. Exploration expenses do, indeed, decline as drilling activity declines. But our definition involves the finding of new wells capable of producing 53 million barrels a day (about 20 billion barrels a year). At that pace, drilling activity would not be so depressed, nor exploration costs as low as they are now. There seems to be no particular reason to change the estimate. From 1980 through 1985 another 100 billion barrels have been used up—not a development conducive to easing the task of finding more oil.

We also should inquire as to the effect of the world recession on short-run oil prices. If it were true that high employment outputs around the world would so raise energy and oil demands as to put the oil industry back near capacity, then short-run prices would be much higher with a full economic recovery. The period of depressed prices could be quite short. Chapter 5 gave some estimates of "high employment" energy use in the Western industrial nations. In looking at the oil component of that demand, it appeared that the recession-induced decrease in oil demand is less than 3 million barrels per day—a small share of the oil industry's excess capacity.[27] It follows, then, that high employment outputs would not now return oil output to capacity and would not pull short-run prices up to $40.

For purposes of considering energy policy, then, we shall use a short-run price of $20 per barrel for petroleum and continue to use $40 as an estimate of the long-run price. This also puts an effective ceiling on natural gas prices, given the competition between these fuels.

Maintaining the pace of energy-efficiency gains and bringing new renewable supplies steadily on the market will be much different in a world of $20 oil than in a world of $40 oil.

Though we have not treated it yet, there is an ever-present risk of another physical supply disruption. The Persian Gulf area supplies 10 million barrels a day at its present reduced rates. There is not nearly enough excess capacity in the rest of the world to offset a disruption of this magnitude. In the event of a major supply shortfall, prices could rise very rapidly.

The overall oil prospect, then, is one of a prolonged period of short-run prices well below long-run price levels, but with the possibility of an oil cutoff and another price jump.

Energy policy must deal with each of these conditions. U.S. energy policy also must reckon with the prospect of a declining output of oil in the United States well before output declines in the rest of the world. These policies are the subject of chapter 9.

Summary

This chapter began by examining the possibility of a high-income world supplied entirely by renewable energy. With a level population of eight billion, and with energy used efficiently, it showed that energy demands can be met from renewable sources.

But that is a distant possibility. The next question had to do with the possibility of getting there from here. A simple growth path in output and energy use was described, leading to an important conclusion. The world can get to a renewable energy future, and can do so without increasing its reliance on coal and nuclear power. Though the growth sketch was a simple one, all of the parameters had some basis in experience. There is a choice among energy futures; none is "historically inevitable."

Since this growth path depends on the maintenance of efficiency gains at rates experienced only after two successive "oil shocks," we must reassess the prospects for this path in a period of declining oil prices. In order to approach this issue, we examined the nature of short-run and long-run oil price determinants—this in order to estimate the likely range of price fluctuations over the next few years. Natural gas prices tend to follow oil prices, especially in the more competitive gas markets now evolving.

Energy policy issues, discussed next in chapter 9, have to reckon with three aspects of oil markets—long-run prices, short-run prices (presently much lower), and the ever-present possibility of a major supply-disrupting event. In the United States the fourth aspect of oil market considerations is the near certainty of declining oil production in the 1990s, and the related prospect of rising imports, or policies to prevent that outcome.

9

Getting There From Here:

Some Policy Proposals

This chapter on policy deals mainly with the United States, and it is oriented toward the energy-efficient, renewable energy future outlined in chapters 2 and 3. As a kind of counterpoint, it briefly discusses policies that might be appropriate to a coal-synfuels-nuclear future. Issues posed by the difference between the short-run and long-run prices of oil are considered. The U.S. situation with respect to its domestic oil resources also poses particular questions for energy policy. Some attention is paid to policy issues in other nations, especially the developing nations.

It seems clear that a transition to an energy-efficient, renewable energy United States is possible. That transition can proceed rather rapidly, by historical standards. The questions for policy formulation are twofold. One set of issues has to do with specific actions to maintain the momentum already established in conservation-efficiency gains and in renewable energy development. The other set has to do with the appropriate pace of the transition in the light of the current world oil and gas surpluses, and the prospective pace of decline in U.S. oil output.

Purists among economists will argue that no one need decide the pace of the transition. The price system will work out whatever is appropriate, they would maintain and the best policy is no policy. I have some sympathy for that point of view, since past government efforts to decide on an appropriate energy future and to force the pace of that future have been less than totally successful. However, many of the policy measures discussed below may be thought of as devices to make the market work better and faster, and those which nudge the economy in the direction that it would take anyway with proper price signals.

Policy measures are discussed under six headings: (1) those designed to help markets work more effectively by removing obstacles that now exist; (2) those designed further to internalize costs still imposed on outsiders, thus distorting decisions in the energy supply industries; (3) those designed to produce marketlike results when, for various reasons, those results are not likely to be forthcoming; (4) appropriate government roles in research and development; (5) special issues arising from the U.S. domestic oil situation; and (6) issues posed by the prolonged divergence between the short-run and long-run prices for oil and gas.

A reordering of U.S. monetary and fiscal policies to reduce the federal deficit and to reduce real interest rates would have benefits in nearly every conceivable sector of the economy. There would be pronounced benefits in the energy sector, especially for energy conservation investments and for renewable energy projects. This is not an energy-specific policy action, but it is important enough to mention in a discussion of policy.

1. Removal of Barriers to the Operation of Competitive Energy Markets

The general principle under this heading is to have prices that reflect long-run marginal costs under competitive market conditions. Some actions already taken fit into this category. The removal of price controls on petroleum products and the partial decontrol of natural gas prices are examples. The Public Utilities Regulatory Policy Act (PURPA) was a large step forward in fostering competition in the supply of electricity. Other policy actions not yet taken include: removal of special tax treatments of oil and gas supply; removal of investment credits and accelerated depreciation throughout the economy. These have been of particular benefit to the conventional energy supply industries. To be sure, they also have benefited energy conservation and renewable energy investments in business, but not in households. The net effect has been to bias energy decisions heavily toward an enlarged conventional energy supply.

—for utilities, removal of tax-free dividend reinvestments.

—removal of liability limitations on nuclear plant accidents, and the inclusion of privately-determined insurance premiums in the cost of nuclear electricity.

—electricity rate schedules which reflect as best they can the long-run marginal cost of electricity at some point of electricity use. This might be related to usage in excess of averages for each customer class.

Several of these provisions were contained in the initial version of a federal tax reform as proposed by the Department of the Treasury in 1984. Unfortunately, these features were largely eliminated in the version of tax reform that emerged in the president's proposals to the Congress.[1] The outcome is not known as this is written.

Price controls were viewed for some time as a means of protecting low-income persons from the impact of higher energy prices. Economists were quick to point out the inefficiency of this attempt. If 80 percent of the population is not poor and 20 percent is poor, then price controls will hold down prices for everyone and encourage overuse of energy. This approach was responsible for part of the rise in U.S. imports after 1974. The method preferred by economists is to make grants to the poor and let prices rise for everyone. The usual policy compromise was reached; prices rose but the grants to the poor got lost in a round of budget cuts.

2. Internalizing External Costs

A number of policy measures have moved in this direction in recent years. Pollution controls in general fit this category, since most air pollution is the result of the combustion of fossil fuels. New power plants must have scrubbers or other pollution control equipment. The costs, appropriately, are reflected in the price of electricity. Much remains to be done. Older power plants and industrial facilities are apparently responsible for the rising damage from acid precipitation. Nuclear accident costs, already mentioned, could be considered under this category of external costs as well.

The external cost that is perhaps the most serious is not easily dealt with under existing institutional arrangements. I refer to the buildup of carbon dioxide in the atmosphere, which could result in heavy damage to agriculture in the United States and to coastal areas if ice caps melt and the oceans rise. Some recent government studies seem to suggest that these developments are inevitable and that the remedy is to prepare for them.[2]

These studies at least have the merit of calling attention to a major problem. Still more alarms have appeared recently, yet the public does not seem to understand.

The chief fault in most government and private studies of the carbon dioxide problem is their failure to consider the alternatives laid out in this book. Climate change, desertification of prime farmland, flooding of coastal cities, and the loss of large parts of Florida and Louisiana—these are serious matters. These problems can be avoided, and they can be avoided without recourse to nuclear energy. Amory Lovins pointed this out in 1981.[3]

Some nuclear advocates use this issue as an urgent argument for nuclear power. The whole message of this book is that neither massive coal use nor nuclear power is essential. Nuclear advocates overlook the equally unpleasant externality of nuclear weapons proliferation; supporters argue that this problem is exaggerated, or will happen anyway. This particular controversy has been well treated elsewhere and is not pursued here.[4] With respect to air pollution from vehicle emissions, there are apparently some advantages from the use of alcohol fuels.[5] Higher-mileage vehicles, regardless of fuel, have the advantage of simply burning less fuel per mile; more efficiency usually means more nearly complete combustion, with less pollution.

These are but examples; the principle is clear—include the costs in the product price so that market choices are not biased.

3. Poorly or Slowly Functioning Markets: The Case for Regulation

There are some situations in which energy-conserving investments or renewable energy investments are economically viable but are not made and might not ever be made. Investment decisions are made by one party, but energy bills are paid by another.[6] People may not know that efficiency gains are possible or how to achieve them. Everyone does not have access to capital for such investments on equal terms. Buildings are long-lived, but the tenancy of any given occupant may be relatively short.

These situations arise most frequently in the residential and small commercial sectors and with respect to energy use for heating, cooling, lighting, and appliances.

The strongest case for regulations, codes, and standards can be made in this area. Builders, developers, and appliance makers already live in a world of codes, standards, and specifications. When they object to energy-efficiency regulations in the name of "free markets," it is chiefly because change is being mandated and adapting to change can be costly. Once new ways of doing things are customary, new standards do not seem to be any more bothersome than the old ones.

In rapidly growing sections of the nation there already is discussion, and some charging, of "impact fees"—for water and sewer connections, roads, and other infrastructure. Energy impact fees have been proposed in Florida. They could be still more effective if they carried generous credits for energy-efficient design. This would help to overcome the "first cost vs. operating cost" problem. It could lead to much more energy-efficient structures and, for that matter, energy-efficient land use for transportation in new developments.

Increasing the energy efficiency of existing structures is another matter. Structures are long-lived and construction practices reflect lags between past practices and those warranted by higher energy costs. Old structures clearly waste energy with respect to present prices; recently built ones do too, and they will be around longer.

Retrofits are more expensive per unit of energy saved than installations in new construction. More research is needed as to the cost and returns of various actions in many different building types and in many different locations. Enough is now known to warrant continuation of insulation, leak-plugging, and other measures to improve the thermal integrity of buildings. When equipment or appliances are ready to replace, the use of high-efficiency models usually pays at present costs of electricity and fuels, without regard to higher, longer-run costs.

Nearly every building undergoes a major renovation at least once in its lifetime. This presents an opportunity for energy-efficiency upgrading similar to that available at the time of new construction. Building permits usually are required, and codes already require such things as wiring upgrading; adding energy-efficiency requirements to codes is no great extension of existing arrangements.

A special case: utility conservation programs.   The theoretical basis for utility involvement in conservation programs for residences and commercial buildings arises from the average-cost-marginal-cost discrepancy in electricity pricing. Consumers see present rates (and, with rare insight, expected future increases). Utilities, on the other hand, directly confront long-run marginal costs, which are higher than present averages, as they consider new generating capacity. Rate reform along the lines mentioned above would make these higher long-run marginal costs visible to customers as well.

There are practical reasons as well for utility involvement in residential and small commercial conservation programs. Utility conservation divisions can develop the body of expertise in diagnosis, installation, and follow-up work for conservation measures. They can develop a network of qualified contractors and the experience that can guide homeowners, landlords, tenants, and small businessmen on what to do, by whom to have it done, what it should cost, and to whom to turn if the work is not properly done. Financing can add additional incentives: low-cost or no-cost loans or even grants for energy-saving actions can be beneficial to all customers, not just the ones receiving the service. This possibility arises from the relatively low cost of saving energy as compared to the rather high cost of building new generating facilities.

For larger commercial and industrial enterprises, energy service companies are being established. They assess energy-saving potentials, install the necessary equipment and controls, and are compensated by a share in energy savings or fees paid from lower energy bills. Some utilities apparently see these as useful avenues of diversification in a changing industry.

Regulation: vehicle mileage standards.   In another sector, auto mileage standards have, until 1984, forced the pace of auto fuel-efficiency gains. In 1985 Ford and General Motors were successful in avoiding fines for noncompliance. Chrysler, which had met the 1985 standards, argued unsuccessfully that the rules should be maintained and that the other two companies could have met the standards with available technology and the same sales mix. The president of Chrysler is presumably well informed in these matters. He has long experience in the industry and worked for many years at Ford.

The retention and perhaps tightening of auto mileage standards largely will determine the pace of continuing efficiency gains, at least for petroleum. Since 1973, auto mileage gains account for about one-seventh of all energy-efficiency gains in the U.S. economy. Petroleum use is increasingly concentrated in transportation; it is becoming a minor factor in electricity generation in the United States and is losing market shares in industry and buildings. A worldwide oil surplus, a public no longer concerned about energy problems, and an administration opposed to regulation—these are present facts. Yet U.S. oil output is certain to decline by 20 to 40 percent in the next two decades. Continued improvement in auto fuel efficiency is central to continuing the transition now under way and to avoiding a surge of oil imports in the 1990s. How do those concerned about energy futures affect policy in the environment of the mid-1980s?

What would they propose? First, a further increase in mileage standards, with standards for light trucks as well. The Audubon Energy Plan proposes a schedule for autos and light trucks that would bring efficiencies to 48 and 42 miles per gallon, respectively, by the year 2000. These are road performances, not simulations.[7] These are reasonable targets; they are technically achievable and economically attractive. Accompanying and reinforcing policy measures include higher gasoline taxes and oil import fees. Other Western democracies have been able to attain average mileages of 30–35 miles per gallon, mainly through very high gasoline taxes and vehicle regulation. This is not directly translatable to the U.S. experience, since these arrangements date from the reconstruction era following World War II. Yet elected governments can receive popular support for such measures. Thirty mpg is far short of the 75 mgp used in chapter 2, but it is higher than the present U.S. standard and much higher than the present average.

The key is leadership and statesmanship. The public will support measures thought to be unpopular when convincing explanations are given and time is allowed for ideas to permeate the public consciousness. Until that kind of leadership emerges in the United States, concerned individuals and groups can try only to point out what lies ahead, and they can have a program ready when another crisis provides an opening for new ideas.

Energy policy and industry. Industrial energy-efficiency gains have proceeded more rapidly than is the case in other sectors. Still, many profitable investments are not made. Corporate capital budget processes apparently have higher thresholds for energy-saving investments than for new product or sales-enhancing investments.[8] Payback times of three years seem to be required, a time horizon similar to that of households.

Industrial efficiency gradually will improve as new equipment is installed. Accelerating the process might entail more generous financial incentives than investment credits alone.

Special problems exist for the most energy-intensive basic industries.

Primary metals and cement are the best American examples. They have aging and inefficient facilities, excess capacity, and strong foreign competition. Their managements tend to ascribe their difficulties to high U.S. labor costs. That is only part of the problem; the overvalued U.S. dollar during the 1980s and unnecessarily high energy costs also are major factors. Profits are low or nonexistent; the financial means to invest in new, efficient equipment are not present.

The case for a "scrap and rebuild" system of grants or low-interest loans can be made.[9] One can devise a system of "competitive reverse bidding" to assure that the financial resources are directed to the firms that will produce the largest efficiency gains for the smallest outlays. There need not be favoritism in the administration of such a program.

Such proposals are not likely to receive serious attention in the present political climate. Some states may be able to assist in programs of this kind. National political climates change however, and ideas rejected in one setting sometimes find acceptance later.

Utility regulation.   Utility regulation offers other opportunities for using regulatory methods to produce marketlike results. Indeed, the time to deregulate electricity production may be near. The technological basis for monopoly in generation—falling long-run marginal cost—no longer obtains with further increases in scale. Industrial cogeneration and a number of small power sources are competitive with central generating stations.

An approximation to deregulation could be accomplished by the observance of the following rule: Any utility wishing to do so would be free to build new generating capacity, providing that it sold itself electricity on exactly the same terms available to any other producer in its service area. Due allowance would have to be made for variables such as reliability, availability during peak periods, or any others that are valid concerns of a utility.[10] Regulatory staffs and commissions can handle issues like these; they have been doing so for years.

4. Research and Development

On the energy efficiency side, there is a need for more information on the efficacy of various measures in buildings of different types and in different climates. This issue is well treated in the *Audubon Energy Plan*. Codes for new construction and retrofit programs for existing buildings need to be grounded in well-verified procedures, and cost-effectiveness needs to be well demonstrated.

All of the estimated gain in appliance efficiency used in chapter 2 can be attained with existing technology. New improvements are possible,[11] and they constitute a valuable research agenda.

In transportation, research in new materials has implications for the next wave of advances in fuel efficiency. Composites with the strength of steel but lighter weight or ceramics for engine use are examples. Aircraft fuel efficiency also is getting some well-deserved research support, but little attention is being paid to improving truck mileage. Rail fuel efficiency likewise receives little attention.

Water transport also is neglected, which is unfortunate since large fuel-efficiency gains are apparently possible as well as the use of a renewable energy source—wind. These are important matters; ocean shipping used 2.5 percent of the world's energy in 1980 and 6 percent of its oil. Coastal and inland shipping use small shares of the national energy totals, but absolute amounts are not trivial. In sail technology Japan is the world leader, with interesting work also going on in West Germany and the United Kingdom.

Industry has been much quicker to respond to energy-saving opportunities than the other sectors, and it also is better positioned to support research in new energy-saving technologies. There are many approaches to increased energy efficiency in industry—new materials, new processes, or new techniques for the recovery of valuable materials that would otherwise be wastes or pollutants. There is no reason to think that new ideas have been exhausted in this area. A system of competitive grants for the development of new techniques should be considered, in addition to basic research that is widely recognized to be a proper activity to be supported by government.

On the energy supply side, chapter 3 listed all of the remaining developments that would be needed to bring the various renewable technologies to technical and economic maturity. Chapter 5 made a direct side-by-side comparison of these and the many remaining tasks to bring a coal-synfuels-nuclear set of technologies to the same degree of maturity.

The long-term outlook for liquid fuels dictates that research attention should be directed to oil substitutes. It already is technically possible to use biomass sources for fuel alcohols and to get liquid fuels from tar sands, shales, and coal. Technology development, operating experience, and cost reduction are needed. This is not basic research, but federal support is nonetheless warranted. There should be two or three small operating units for each of the conversion processes not now commercialized—biomass methanol, and liquid fuel from shales and coal. The object is to learn from actual, but small-scale, operating experience, to improve conversion technologies, and to provide a firm basis for estimating costs in larger-scale facilities.

In the renewable energy area, ethanol is already commercialized with generous tax subsidies, and cost-reducing technological advance is proceeding. Methanol conversion is held back by surplus capacity in fossil-fuel methanol sources. Yet we need to know more about costs and technologies in biomass methanol.

Tar sands have been converted into petroleum products in Canada since 1967. Attempts to scale up this process and to reduce environmental impacts are proceeding. Shale oil projects have been scaled up too rapidly, for the process as yet does not seem to work.[12] Coal conversion to liquid fuels lags in the United States. All of these efforts are influenced by falling oil prices. The long-run outlook for liquid fuels has not changed; it is just that the long run is a little longer.

A well-ordered research program in biomass-to-methane processes is already under way. It is comprehensive in scope, ranging from high-yielding feedstock species to harvesting, preparation, and conversion technologies. The natural gas industry through the Gas Research Institute has been more farsighted than the petroleum industry. Even here, the short-run oversupply of natural gas has resulted in research cutbacks. This program could well qualify for federal support.

In other renewable energy technologies, photovoltaics and solar industrial process heat can benefit from basic research in materials. Market development and operating experience are now encouraged by renewable energy tax credits, which are expiring. Financial incentives to continue these developments will be needed. Again, the object is to learn from experience and to maintain a steady progression toward market penetration without further subsidies. Windpower is nearly there; tax credits seem to have operated more effectively in creating actual operating windpower capacity than did direct federal expenditures for large experimental wind machines.

There are new technologies in the renewable energy area—Ocean Thermal Energy Conversion, for example—which warrant continued investigation.

The most important issue in federal energy research and development is evenhandedness. Research budgets have been badly skewed toward nuclear electricity, with little attention to liquid fuels, heat, and other electricity sources. In the absence of a better rule, three roughly equal shares—conservation/efficiency, renewable energy, and nuclear-coal-synfuels—could serve as an allocation rule for federal energy expenditures. The size of the total budget need not be much different from that of recent years. It has been cut in an administration dedicated to reducing the size of the federal government, but the Congress has successfully resisted the most drastic cuts in renewable energy research.

There are energy problems ahead; devoting 1 percent or so of the nation's energy expenditure to energy research would not fare badly in a national referendum. The issue tends to get lost in political battles over larger sums and in the present doctrinaire approaches that seek to shrink all government activity.

5. Implications of Short-run Oil Prices

All of the discussion of policy so far except research and development has been market-oriented—removing obstacles, internalizing costs, and using regulation to produce marketlike results. The next question to be considered is the extent to which this program is valid when oil and gas are in surplus, and prices for them are well below long-run costs. Should policy aim for less when oil is $20 a barrel instead of $40 a barrel? It would seem that all actions that assist markets to work more effectively, or that produce marketlike results, would be warranted in any case. Even with conflicting signals from oil and gas prices, electricity prices are likely to continue moving toward long-run prices. Implementation of the policies affecting electricity pricing would assure this result. All of the conservation-efficiency actions regarding electricity outlined in chapter 2 still would be economical.

It is in transportation and industrial fuel markets that policy might be most affected by the situation of the oil market. Consideration of these matters is so closely tied to the next issue—U.S. domestic oil and gas production—that the two issues are treated together.

6. The U.S. Oil Situation

If the United States maintains the pace of efficiency gains achieved in the period from 1979 to 1984, energy use would decline slowly over the next several decades. Overall rates of energy-efficiency gain would exceed rates of economic growth so that there would be a gradual absolute decline in energy use—.5–1 percent per year. The development of new sources of renewable energy along the lines discussed in chapter 8 would further decrease the use of fossil fuels. In fact, this time path for economic growth, energy-efficiency gains, and renewable energy development would accommodate U.S. economic growth and the probable time path of oil output decline as well.[13] It would be consistent with a policy aiming to keep oil imports at approximately their 1985 levels. An analysis of overall energy totals is not sufficient. One also must look at end uses and energy forms. The demand for petroleum must decline at the appropriate rate. Petroleum use is now so concentrated in the transportation sector that higher auto mileage standards probably are essential to staying on the path just outlined. There may be other ways, but they are not readily visible from the data.

The auto-buying public and the auto manufacturers see, in 1986, gasoline at slightly under $1 per gallon, falling world oil prices, and a large surplus oil production capacity. As countless energy analysts have warned, the long-term realities are rising oil imports, another price jump, or perhaps a sudden supply interruption.

Should auto mileage increase a decade or two ahead of market signals as they now appear? Should gasoline taxes, petroleum product taxes, or oil import fees be used to reinforce mileage standards? Will the public support them? These are the chief energy policy issues of the moment.

From a national security point of view, or from a full marginal cost-of-imports point of view, or from a long-run view of the national interest, there is a strong case for one or more of the policy measures just listed.[14]

As economists view resource allocation decisions, it would seem appropriate to import relatively cheap oil if that is what buyers want to do. When costs are fully stated, this conclusion may not follow. There is a choice; an attainable conservation-efficiency path exists and the policies to stay on it are known.

Presidents are best situated to exert leadership—to lay choices before the public, to propose a course of action, and to explain the reasons for it. That road to policy formation for petroleum is not presently available in the United States. States can take some initiatives but not easily on matters such as auto mileage standards. For now, to repeat a point made earlier, concerned groups and individuals can try to keep these issues under public discussion. A supply disruption in the Persian Gulf, always a possibility, would quickly refocus attention on oil import vulnerability. Keeping the issue alive, with a view of the problem and proposed solutions, is the best that can be done. Someday, new leadership will emerge, and new ideas will have their opportunity.

In other areas, policies for the transition can move ahead. This chapter has suggested about forty different policy actions, many of them independent of the others. Many are applicable at any price of oil in the $20 to $40 range. Some are partially in place. A roadblock in one policy area does not mean that all of the others also are blocked. That logic explains the present drive for state-imposed appliance efficiency standards.[15] The federal Department of Energy refused to develop and enforce them from 1981 until 1985. It has, in 1985, been ordered by the courts to do so. It successfully may delay a few more years, but standards gradually are being imposed by the states anyway. State utility commissions are moving to implement some of the measures described here. Electricity conservation has every reason to proceed, and renewable electricity gradually works its way into electricity markets. Biomass-based methane inches toward commercial status. Biomass-based liquid fuels and, for that matter, liquid fuels from shale, tar sands, and coal are likely to be the main casualties of oil market conditions unless policy incentives are put in place.

In summary, U.S. policies to maintain the present momentum in efficiency gains and in renewable development would be adequate to keep oil imports from rising. They would have to include further tightening of auto efficiency standards. This particular development seems, in 1986, to be unlikely, but many other gains are possible for those holding views similar to the one expressed here. Times and leaders change; one must be prepared with the ideas that will later find a hearing.

An Aside: Policies for a Coal-Nuclear-Synfuels-Fuels Factor

This chapter so far has focused mainly on policies that would be appropriate for an energy-conserving, renewable energy future. It is instructive to look also at policies appropriate to a high-energy, high-electric, high-nuclear future, which also relies heavily on coal and synthetic fuels. A modified view like this is still the "establishment" view, so it must be accorded some attention.

A beginning point is a look at 1985 figures, and to ask what set of policies, consistently applied since, say, 1975, would have produced an outcome like those projected in the mid-1970s. This approach also illuminates the extent to which the energy debate has moved to new territory over the decade.

A typical mid-1970s projection of 1985 energy use was quoted in chapter 5. Total energy use was expected to be 105 quads. Oil, gas, coal, and nuclear power were expected to contribute, respectively, 42, 26, 13, and 19 quads. Actual 1985 use was about 74 quads.

The task now is to see what set of policies might have produced results more like the earlier projections. Part of the answer is in real energy prices. Most 1970s projections were explicit only about the price of oil, if that. Using standard elasticities, it appears that subsidies of $150 to $225 billion[16] would have been required to induce a level of energy use so much larger than the actual use, or an estimated "high employment" use. More accurately, additional subsidies of $150 to $225 billion would have been required, since existing subsidies have been estimated at $50 billion.[17] This estimate is probably on the low side, since higher oil use probably would have led to still higher oil prices, thus requiring still larger subsidies.

People who made these projections in the 1970s did not have in mind a world of high and rising energy subsidies just to maintain high levels of energy consumption. They expected that more domestic oil would be forthcoming, that nuclear-generating costs would fall (and thus maintain the historic decline in real electricity prices), and that technologies in coal combustion and synthetic fuels would develop more rapidly. In considering policies to foster this kind of energy future, we can start by asking what policies, other than subsidies, were not applied, since the mid-1980s are turning out to be so different from expectations. The failure of domestic oil production to rise is apparently a miscalculation of supply elasticities and also of remaining petroleum resources. Both are uncertain, and it is very easy to be wrong in these matters. Nuclear electricity costs rose in real terms when they were expected to fall. The industry cites regulatory inadequacy and interest rates. Yet rising real costs have been experienced in every other nation developing nuclear power, without regard to its regulatory system. Real interest rates have risen over this period in all nations, though not as much as in the United States. France is the nation in which the nuclear program has proceeded without hindrance, at least until 1984; real costs there at the margin are now higher than the U.S. average. France provides no evidence that the real price of nuclear electricity can fall over time, either there or here.

There has not been a systematic and well-articulated research program in coal-combustion technologies and in synthetic fuels. This is an area in which some progress might have been made in technology refinement and in cost reduction.

As one looks ahead, one can ask whether there is any set of policies. other than massive subsidies, that is likely to produce a high-energy, high-electricity, high-nuclear future? Rapid development of technology would have to lower nuclear-generating costs, complete the fuel cycle at much lower costs than now exist anywhere in the world, and develop safe, reliable, low-cost breeder reactors. Synthetic fuels processes capable of producing 30 to 40 million barrels a day would have to be developed, with costs of liquid fuel outputs in the $1 to $2 per gallon range. Developments such as these are imaginable in the strict sense of the word. They do not seem all that likely, however. The mere listing serves to underline the modesty of the requirements for an energy-efficient, renewable energy future, as contrasted with the requirements for a coal-nuclear-synfuels future.

U.S. Policy Summary

Among the many policy actions that might be taken, the following seem to be the most important:

—The ending of subsidies to conventional energy sources. The extension of subsidies to conservation and renewable energy investments is only second-best. Energy overuse is encouraged. This is a federal government matter.

—The extension and increase of vehicle mileage standards—also a federal government matter.

—Tightening of building codes with respect to energy use—a state and local government matter.

—Appliance efficiency standards with progressively more stringent requirements. Preferably a federal matter, but one proceeding now at the state level.

—A balanced research effort with respect to conventional energy sources, energy conservation, and renewable energy technologies.

These measures, and many of the others already mentioned, probably would maintain the pace of energy-efficiency gains already experienced and maintain the pace of renewable energy development. The world oil market surplus, and thus the divergence between short- and long-run oil prices, will probably end in the 1990s. If the momentum in conservation and renewable energy development can be maintained over that period, the nation will be well along toward the energy future described here.

Policy Abroad

Policy in other industrial nations and in Third World nations is their own business. That does not mean it is beyond U.S. influence. Indeed, the U.S. influences decisions in other nations whether it tries to or not, or whether it does so consciously or not.

The adoption of a conservation-renewable energy future in the United States would have an incalculable influence on the rest of the world. This would have to be more than a presidential announcement. Presidential proposals, implementing legislation, and visible progress toward a declared goal would all be essential elements. These actions would dramatically change the terms of discussion of energy policy in every nation of the world. The United States need not urge anyone to do anything—a temptation to which it too often succumbs. The power of its example would be more than enough.

Should developing nations choose to embark on similar energy development paths, it might be helpful to review some appropriate policy actions. Theoretical feasibility is one thing; concrete action to follow a renewable energy, energy-efficient path is another.

Energy policies are inextricably linked up with the overall development process. This is a vast field with a voluminous literature. Perhaps the best thing that can be done here is to note the ways in which the development process would be affected by a conscious move toward a renewable energy future.

Some clues can be found in the experience of the industrial nations. There are no high-income nations, excepting the special case of Saudi Arabia, that do not have productive agricultural sectors. Western Europe is now self-sufficient in food, and Japanese agriculture is very productive as well. North America, Australia, and New Zealand are longstanding exporters of agricultural products. In these same areas, forest cover actually has increased in the last several decades despite some population growth and urban spread. Agriculture and forestry are rather small sectors in terms of employment or national product, but that reflects their high productivity, not their lack of importance.

These two sectors are vital for the development of renewable energy, especially biomass. Forests also protect watersheds for hydroelectricity. The importance of agriculture and, to some extent, forestry has been receiving increasing recognition in the literature and in the practice of development. Rural sectors frequently were neglected in development planning and budgets. Industry and urban areas were favored. Price policies were biased against farmers and toward urban populations. These policies increasingly are recognized as having been mistaken by the international institutions and by developing nations themselves.

This concern with agriculture and forestry needs only a modest extension to promote the development of renewable energy as well. The opportunities need to be recognized and integrated into development projects.

Industrial energy efficiency needs more attention in the international institutions and in national development banks. Project development and evaluation need an explicit energy component just as industrialized nations have energy managers in their firms. The knowledge is available in the world. What is needed is management recognition of the high potential returns to investment and the presence of staff members with competence in industrial energy use. Nearly every developing nation of any size has an industrial development bank. Many are supported by the World Bank and the regional development banks. The institutions are in place; it is a question of working the knowledge and the staff people into them.

Investments to increase industrial energy efficiency should be as readily financed as any other industrial project with similar prospective returns. New equipment should be screened for energy efficiency as well as for other technical and economic factors.

Electricity supply has long been an area of support by the international institutions. What is needed is a new awareness in the national electric enterprises and the lending institutions. Hydroelectric development has been particularly rapid in developing nations and has received heavy support from the World Bank and the regional banks. Geothermal sources are beginning to get recognition in the same way. Small hydro-electric installation and rural electricity supply need more attention.

Wind potential surveys are natural candidates for support in multi-lateral or bilateral aid programs. The costs are relatively low, and the returns potentially quite high, as California's experience indicates. Developing nations already have shown a great deal of interest in photovoltaic cells; they will be among the many beneficiaries of falling array prices.

For the semiarid African and Asian nations, the development of solar ponds is a neglected matter. This technology can provide base-load power in areas where hydroelectric sources are scarce.

Efficient building design is quite site-specific. Much of the work done in the colder areas of Europe and the United States is not transferable to developing nations. Even in the United States, too little is known about building efficiency.

Those who attain income levels at which building indoor comfort is bought frequently buy that comfort inefficiently. Regional research centers in each of the several climate zones in Africa, Asia, and Latin America should command support from the developing nations themselves as well as from the international agencies.

Vehicle efficiency standards are more easily imposed if there also are standards in the industrial nations. Vehicle markets (autos, trucks, aircraft) are increasingly international in scope.

One might ask how all of these things can be done when half of the developing world is struggling to avoid slipping backward. Latin American growth is stymied by large foreign debts, and half of Africa faces institutional collapse in the face of multiple challenges. Perhaps development cannot continue, or it cannot continue everywhere.

One also must consider other parts of the developing world. China reportedly has doubled rural real incomes in six years and its national product in ten years. Its rate of population growth is slowing down. India is now self-sufficient in food and maintains steady growth in per capita income. Indonesia, the world's third largest developing nation, doubled its national product in the 1973–83 decade. All sectors, not just oil, grew rapidly. Still, Indonesia's population grows at rates that will lead to alarming numbers in a generation, as does India's.

Ten sub-Saharan nations of Africa with 15 percent of its population have continued to grow rapidly despite the crises that afflict the rest of the area. Population growth rates, though, are among the highest in the world.

The developing world, as usual, presents a diversity of situations. One can be very optimistic or very pessimistic.

All that can be said here is that, if these problems have solutions, energy efficiency and renewable energy have a role. If the problems do not have solutions, it is not the energy components that stand in the way.

Again, the most important single contribution that the United States can make is that of example: to set out on the energy-efficient, renewable energy path deliberately and consciously.

The irony of the matter is that we are already launched on such a course. If one takes studies like SERI and the Audubon Energy Plan, estimates for energy conservation and renewable energy by the year 2000 are almost linear extrapolations of U.S. energy history from 1979 to 1985. The huge missing element is the understanding that the road we have happened on by accident is one that leads to a desirable and attainable energy future. Without some understanding, some efforts, and some positive policy actions, we will wander off that road. Getting back on it will turn out to be much more costly than simply realizing where we are and staying there.

Notes

1 The Energy Debate: More Divergence than Consensus

1. The views on nuclear power of the early and mid-1970s involved quite rapid growth—perhaps 1,000 reactors operating by the year 2000. This view of the nation's energy future implied a very large share for electricity as oil and gas supplies diminished. Space heat, some industrial process heat, and perhaps (with batteries) some vehicle fuel would be supplied by electricity. This prospective growth raised questions about the adequacy of uranium supplies. Plans therefore called for the extraction of still-fissionable uranium from spent fuel rods, and plutonium, which forms in the operation of reactors, as well. This "reprocessing" of spent fuel would extend uranium supplies and provide plutonium for "breeder" reactors, which also are necessary to a high-nuclear future. Chapter 5 deals with these matters.

2. As contrasted with coal, oil, or natural gas, which are physically destroyed by combustion and, on time scales of human interest, are not renewed.

3. Energy Research and Development Administration, *A National Plan for Energy Research, Development and Demonstration ERDA 48* (Washington, D.C., 1975); International Energy Agency, *World Energy Outlook* (Paris: OECD, 1977).

4. H. J. Barnett and C. Morse, *Scarcity and Growth: The Economics of Natural Resource Availability* (Baltimore: Johns Hopkins University Press for Resources for the Future, 1963).

5. See, for example, Roberl M. Solow, "Is the End of the World at Hand," *Challenge*, March–April 1973. Solow is taking issue with the resource pessimism in the Club of Rome's *Limits to Growth*. Goeller and Weinberg refer to energy as the ultimate raw material in an age of "substitutability." *American Economic Review*, December 1978.

6. This view is partially descriptive of the Reagan administration's approach to energy matters. It is held by most economists in the federal agencies. Another strand of administration thinking is evidenced by increased Department of Energy spending on nuclear energy sources, while support for energy conservation and renewable energy is reduced. Another deviation from the "free market" approach is the retention of special

tax benefits for the oil and gas industry, while conservation and renewable energy tax credits are set to expire.

7. Economists associated with the Ford Foundation's third study of energy problems express these views. *Energy: The Next Twenty Years.* Report by a Study Group Sponsored by the Ford Foundation and administered by Resources for the Future (Cambridge, Mass.: Ballinger, 1979).

8. The term is used as a title in John P. Gibbons and William U. Chandler, *Energy: The Conservation Revolution* (New York: Plenum, 1981). This is one of several important works cited in chapter 2.

9. Energy Policy Project of the Ford Foundation, *A Time to Choose* (Cambridge, Mass.: Ballinger, 1974).

10. Morris A. Adelman et al., *No Time to Confuse* (San Francisco: Institute for Contemporary Studies, 1975).

11. *Alternative Energy Demand Futures to 2010.* Report of the Demand and Conservation Panel to the Committee on Nuclear and Alternative Energy Systems (CONAES), National Research Council (Washington, D.C.: National Academy of Sciences, 1979).

12. Earlier versions, with higher rates of growth in energy use and with high coal and high nuclear components, are cited in note 3 above. More recent versions start at lower levels of energy use and have lower growth rates. Still, they lead to similar energy futures with a twenty- to thirty-year lag. See, for example, *Annual Energy Outlook 1984* (Washington, D.C.: Department of Energy, 1985). Total energy use in the United States was projected to grow from 73.7 quads in 1984 to about 90 quads in 1995. Coal and nuclear power provide much of the increase of 16 quads. Oil use is projected to rise but only about 4.5 quads. For the industrialized nations, the International Energy Agency made a similar projection in 1982. Any model in which economic growth proceeds, energy conservation is modest, energy growth is exponential, and the oil-gas contribution levels off will produce a similar result. The IEA's projected demand for the year 2000 was about 40 percent lower in 1982 than it was in 1977, but the increasing reliance on coal and nuclear power present in the earlier projection is still an important feature. International Energy Agency, *World Energy Outlook* (Paris: Organization for Economic Cooperation and Development, 1982). The IEA repeats essentially the same message in its 1985 update.

13. Amory Lovins, "Energy Strategy: The Road Not Taken," *Foreign Affairs*, October 1976.

14. Amory Lovins, *Soft Energy Paths* (San Francisco: Friends of the Earth, 1977).

15. Denis Hayes, *Rays of Hope: The Transition to a Post-Petroleum World* (New York: Norton, 1977).

16. Henry W. Kendall and Steven J. Nadis, eds., *Energy Strategies: Toward a Solar Future* (Cambridge, Mass.: Ballinger, 1980).

17. Bent Sorensen, *An American Energy Future* (Copenhagen: Niels Bohr Institute, University of Copenhagen, 1980). A more accessible version is in *Energy*, vol. 7 (1982).

18. John S. Steinhart et al., *Pathway to Energy Sufficiency* (San Francisco: Friends of the Earth, 1979).

19. Robert Stobaugh and Daniel Yergin, eds., *Energy Future.* Report of the Energy Project at the Harvard Business School (New York: Random House, 1979).

20. Daniel Deudney and Christopher Flavin, *Renewable Energy—The Power to Choose* (New York: Norton, 1983).

21. It is customary in these matters, especially for economists, to try to show that something is "cost-effective," since anything that is not is not worth talking about. Hence the showing, in this study, that conservation-efficiency and renewable energy are "cost effective." This approach carries the presumption that a free-enterprise market economy will move in these directions without further ado. The history of the energy industries in the United States suggests that reality is not quite so simple. For a detailed examination of these issues, see Craufurd D. Goodwin, ed., *Energy Policy in Perspective* (Washington, D.C.: The Brookings Institution, 1981).

2   The U.S. Energy Conservation-Efficiency Potential

1. The year 1980 might seem to be a questionable choice for the analysis; a more recent year might seem preferable. Oil and gas prices have been declining recently and are now lower in real terms than in 1980. There are several reasons for the choice. One should not talk about conservation potentials or savings percentages except in reference to clearly specified figures. For that purpose, data from any recent year would do. The year 1973 might be better in a way, for no response to energy price increases had yet begun; one thus would avoid double counting of savings. Yet 1973 is rather far back for a study concluded in 1986. Some response to price changes had taken place by 1980, but not much, since oil and gas prices were controlled and the second wave of price changes was just then taking place. Nineteen eighty is the latest year for which good end-use data are readily available, and those data are critically important.

   Since real prices for oil and gas are now lower than in 1980, is there not some danger of estimating too large a conservation-efficiency potential? Perhaps so, for short-run purposes, but probably not for long-run purposes. This issue is addressed in chapter 9.

2. Total primary energy use was 76 quads. Sixteen quads were lost in producing electricity. More than 3 quads were used up in oil refining, and 1 quad of natural gas was used in "mining"—that is, in getting oil out of the ground. Another quad or so was used in various ways in the energy sector—enriching uranium, transporting pipeline fuels, and the like. Three quads are for nonenergy uses—feedstocks and asphalt.

3. This point was made early on by Amory Lovins and was a key element in his critique of high-nuclear, high-electricity energy supply scenarios. Lovins, *Soft Energy Paths*.

4. This notion of comparing investment returns has been used in several recent studies. The idea is to compare (*a*) investments in conventional energy supply expansion, (*b*) investments in efficiency enhancements, and (*c*) investments in renewable energy sources, and then select those with the highest returns. One of the great discoveries of the last decade is that efficiency investments win by a large margin—a point developed much more fully in the pages ahead. This approach was first popularized by Roger W. Sant, *The Least-Cost Energy Strategy* (Arlington, Va.: Energy Productivity Center, Mellon Institute, 1979), and John C. Sawhill, *Energy Conservation and Public Policy* (Englewood Cliffs, N.J.: Prentice-Hall, 1979), p. 6.

   Most energy experts of the 1970s, drawn from the energy supply industries, tended to define the problem as one of somehow, with enormous effort, increasing supplies to meet very large projected future demands, the latter being somehow given and beyond the reach of analysis. Most of them still define the problem this way, but with smaller numbers out in the future. For those in the supplying industries who would not read Sant or Sawhill, the lesson has been brought home by the "invisible hand," or as it may appear, the "iron fist."

5. This example suggests another important point: there is much room for further improvement in refrigerator efficiency in the United States if designers and manufacturers have stopped that far away from the margin with their best models. This situation will be encountered repeatedly as other uses of energy are examined.

6. In chapter 3 renewable energy supplies will be considered. The "investment portfolio" approach will be broadened to compare investment returns in conventional energy supply, energy conservation-efficiency, and renewable energy supply.

7. IEA, *World Energy Outlook* (1982); Energy Information Administration, *International Energy Annual, 1983* (Washington, D.C., 1984).

8. Boum-Jong Choe, *A Model of World Energy Markets and OPEC Pricing.* World Bank Staff Working Papers Number 633 (Washington, D.C.: World Bank, 1984). The $42.50 price guarantee is for Union Oil Company's shale project and is reported in the company's quarterly report to shareholders, April 1983.

9. World Bank, *The Energy Transition in Developing Countries* (Washington, D.C., 1983).

10. Costs to U.S. oil companies for adding reserves were reported at $15 a barrel in 1981 and $17 in 1982 (Arthur Anderson study reported in the *Wall Street Journal*, January 10, 1984). This is still not quite the figure we seek. After discovery of new reserves, other costs, which must be covered, are entailed in extracting oil ready for sale. Also, our figure represents the cost of finding oil in quantities equal to the quantities being used up. U.S. discoveries have not attained that rate in most of the past fifteen years. A *Wall Street Journal* story (February 14, 1985) quotes various oil industry officials on the relationship between oil prices and exploration activity. Those views are reflected in the discussion in the text.

11. The Gas Research Institute uses a figure of $5–$6 (1983 dollars) as a competitive gas price for the year 2000. My figure ($3.50–$4.00 in 1980 dollars, adjusted to $4.25–$5.00 in 1983 dollars) is an estimate of the price now of finding new gas reserves and delivering gas to the consumer. The GRI figure, which is higher, represents the cost of finding gas in the 1990s for delivery around the year 2000. Gas Research Institute, *Methane from Biomass and Waste* (Chicago, 1984).

12. One kilowatt hour provides 3,413 btu's; to buy 1 million btu's, one must buy 293 kilowatt hours. At 12¢ per kilowatt hour this costs a bit more than $35. One barrel of oil contains about 5,800,000 btu's, and 5.8 times $35 is $203. This point was made by Lovins, *Soft Energy Paths;* and by Vince Taylor, *Energy: The Easy Path* (Cambridge, Mass.: Union of Concerned Scientists, 1979).

13. Stated in 1980 dollars for comparability with other 1980 cost estimates. These figures are from the CONAES study. The National Research Council, *Energy in Transition 1985–2010.* Final Report of the Committee on Nuclear and Alternative Energy Systems (San Francisco: Freeman, 1979), p. 139. These estimates are now regarded as too low.

14. For example, fluidized bed combustion captures sulfur compounds in the "bed" and thus avoids costly stack scrubbers. In electricity generation, efficiencies of 40 percent are attained as contrasted with 35 percent in the best presently operating plants. This is

probably the most promising technology for coal industrial heat as well as for new coal-fired generating plants. Since industrial fluidized bed units can be cogenerators, competition in electricity supply will be enlarged.

15. IEA, *World Energy Outlook.* (1982).

16. One recent report based on careful monitoring is found in *Solar Age*. This is a summary of a report by Douglas Balcomb, solar energy advisor at Los Alamos and a U.S. solar pioneer. The best performer, a house in Colorado, used auxiliary heat amounting to less than .5 btu per degree day Fahrenheit per square foot. *Solar Age*, April 1983, p. 56. The 1980 U.S. average was about 20. See chap. 2, n. 17.

    Energy use per household has been declining steadily, especially since 1978—sufficiently rapidly to offset the increase in the number of households. Most of this, however, represents life-style changes (people willing to live with colder room temperatures in the winter) rather than conservation-efficiency as the terms are used in this study. The long-run potential gain, of course, is still there. These matters are examined in *Technology Review*, October 1983.

    A recent data summary from Lawrence Berkeley Lab's Building Energy Data Group showed that low-energy houses saved energy at costs that averaged 2¢ per kilowatt hour or, in gas-heated homes, $4.40 per million btu's. Reported in *Solar Age*, August 1985, p. 45.

    Prefabricated houses built to Swedish standards and with some imported Swedish components will, if they match Swedish performance, use .5 btu per square foot per degree day—about 2 percent of the U.S. 1980 average. *Solar Age*, April 1985, p. PB6.

17. D. Neeper and R. McFarland, *Some Potential Benefits of Fundamental Research for the Passive Heating and Cooling of Buildings* (Los Alamos, 1982). The average energy use for residential space heating in the mid-1970s was about 22 btu's per square foot per degree day Fahrenheit. A 1,500-square-foot residence in Washington, D.C. (4,000 heating degree days), would take 139,000,000 btu's, or 139,000 cubic feet of gas. Construction practices in 1980 had improved so that new houses averaged about 10 btu's per square foot per degree day. New construction had thus reduced heating demand, for the average house, to about 20 btu's (data from SERI, *A New Prosperity*, pp. 50–61). I am asserting that this average can be reduced to 1–2 btu's per square foot per degree day.

18. Researchers at the Florida Solar Energy Center are developing a passive cooling/dehumidifying technique using a dessicant such as silica gel. The dessicant is located in the attic, which is thermally isolated from the living space by heavy insulation. The house uses block walls, as do nearly all Florida houses, but with exterior insulation. Heat and humidity are absorbed during the day by the walls; at night fans move heat and humid air to the attic where the dessicant cools and dehumidifies the air. The absorbed heat warms the roof, which then more readily radiates heat into the night sky. During the day the attic heats from the sun, and blowers remove the humid air as the dessicant recharges. Florida Solar Energy Center, *The Solar Collector* (Winter 1985).

    This development is not taken into account in my calculations; it is cited as an instance of continuing research that may add to the conservation-renewable energy potential in the future.

19. One set of techniques for reducing cooling loads is described in *Popular Science*, April

1985. Masonry walls with exterior insulation are used to put additional thermal mass inside the insulated space. Reflecting materials under the roof but above the ceiling insulation reject heat. A small air-conditioning unit, one-third the normal size, cools and dehumidifies. The structures described are for Louisiana and the Texas Gulf Coast; similar techniques are being developed in Florida and are applicable to much of the Southeast.

20. Lennox, in 1984, offered an air-conditioner with an energy efficiency ratio of 15, more than twice as efficient as those now in use. The existing average for central units is 6–7, and newly installed units are typically in the 7–9 range. SERI, *A New Prosperity: Building a Sustainable Energy Future* (Andover, Mass.: Brick House, 1981), pp. 65 and 77.

21. Saved electricity from high-efficiency air-conditioners costs about 4.5–5¢ per kilowatt hour by one Florida calculation. Barney L. Capehart, John F. Alexander, and Lynne C. Capehart, *Florida's Electric Future: Building Plentiful Supplies on Conservation*, (Gainesville: University of Florida Center for Wetlands, 1982). In the author's own experience, savings appear to cost less than 3¢ per kilowatt hour (calculated from residential installation, Maitland, Florida, 1984–85).

22. An interesting variation of this problem appears in warm, humid climates. Buildings in which efficiency measures have reduced cooling loads, and in which high-efficiency air-conditioners have been installed, have too much indoor humidity because the air-conditioner does not run enough. One solution is to overcool the air (thus removing moisture) and then reheat it. This procedure increases energy use again. A Florida Solar Energy Center researcher, Mukesh Khattar, uses a heat pipe to cool building air entering the air-conditioning unit; the heat accumulated is transferred by the pipe to overcooled air leaving the unit. This avoids energy use to reheat air and allows a smaller, more efficient unit to cool and to dehumidify. The technique is widely applicable. Mukesh Khattar, paper at Florida Solar Coalition Energy Research Forum, October 1984, Miami Beach. Also reported in *Solar Age*, May 1985.

23. North American Phillips is now advertising a 19-watt bulb with the same illuminating power as a 75-watt incandescent bulb. It costs more but requires fewer changes. For overall gains, see *New Scientist*, March 17, 1983. For one of many references on daylighting, see *Solar Energy Digest*, November 1982.

24. R. R. Verderber, *Lighting Conservation—An Energy Resource?* (Berkeley: Lawrence Berkeley Laboratory, 1981).

25. *Soft Energy Notes*, March–April 1982, p. 26.

26. With the exception of Carrier, which supports standards.

27. A new gas burner, designed to minimize air pollutants (especially nitrogen oxides), has the added property of transferring much more heat to the bottom of the pan than burners now in use. Fuel saving is about 40 percent. *Science News*, January 14, 1984.

28. Water-saving measures alone reduce hot water use some 50 percent (SERI, *A New Prosperity*, p. 69). The other measures easily provide the remaining 10 percentage points.

29. This calculation is shown in the appendix at the end of this chapter.

30. California Energy Commission, *Securing California's Energy Future, 1983 Biennial Report* (Sacramento, 1983), p. 81.

31. Data for the Mount Airy, N.C., Library (actual performance) are found in *Solar Age*, May 1985. The Gulf building in Denver is reviewed in *Solar Age*, July 1984. The New Jersey citation is a Prudential building in Princeton; energy data are given in *Solar Age*, December 1981. For comparison, SERI (*A New Prosperity*, p. 156) gives average commercial building use in 1980 as 18 kwh in electricity and 135,000 btu's in fuel for each square foot of space. Considering the fuel burned to produce electricity, this comes to 342,000 btu's per square foot of commercial space. Construction practice in 1980 for new buildings had reduced this to an average of 205,000. The 1985 California code would bring this to 138,000 btu's. The three buildings cited, again counting electricity on the basis of fuel burned to produce it, used 67,000–79,000 btu's per square foot (all figures for space conditioning and lighting). The California Energy Commission estimates that energy use in commercial buildings can be reduced by 80 percent, which is my estimate. California Energy Commission, *Securing California's Energy Future*.

32. *Monthly Energy Review*, February 1985.

33. *New York Times*, December 22, 1982.

34. *Orlando Sentinel*, September 16, 1985.

35. *Wall Street Journal*, May 29, 1984.

36. *Business Week*, March 12, 1984, p. 91.

37. *New Scientist*, May 9, 1984, p. 177.

38. *New York Times*, November 22, 1981, quoting Sherwood Fawcett. New technologies are under active investigation. Ceramics, which permit higher combustion temperatures, hence more efficient engines, represent one interesting field of exploration. These advances, not included here, suggest that still more gains will materialize in the future.

39. Emmet J. Horton and W. Dale Compton, "Technological Trends in Automobiles," *Science*, August 10, 1984.

40. *Science*, April 19, 1985.

41. Boeing, Quarterly Report to Stockholders, November 1982.

42. E. Hirst et al., "Recent Changes in U.S. Energy Consumption," *Annual Review of Energy, 1983*, Jack M. Hollander, ed. (Palo Alto, Calif.: Annual Reviews, Inc., 1983).

43. Other technologies are under development but are not included in these estimates. "Propfan" engines can move planes at jet speeds with fuel savings of 20–50 percent. General Electric, Boeing, and McDonnell Douglas are the firms involved. Media accounts appeared in the *Orlando Sentinel*, July 21, 1985, and *Popular Science*, March 1985.

44. *Soft Energy Notes*, May–June 1982.

45. The SERI result assumes no gains in rail fuel efficiency. This is not the case, as the following note 46 indicates.

46. From 1972 to 1982 Class I railroads showed a 24 percent increase in fuel efficiency per revenue-ton-mile. Further gains are possible in new equipment through engine efficiency gains, attention to aerodynamics, and improved wheel assemblies. *Technology Review*, February–March 1984.

47. Reported in *Popular Science*, January 1985.

48. New York City Energy Office, *Energy Consumption in New York City* (New York, 1981), p. 64.

49. CONAES, *Alternative Energy Demand Futures to 2010*.

50. Office of Technology Assessment, *Industrial Energy Use* (Washington, D.C.: U.S. Government Printing Office, 1983).

51. Marc H. Ross and Robert H. Williams, *Our Energy: Regaining Control* (New York: McGraw-Hall, 1981), p. 6.

52. Reported in *Physics Today*, August 1975. "Second Law" (of thermodynamics) efficiency refers to energy actually used in a given task as compared with the most effective way of providing that energy.

53. CONAES, *Alternative Energy Demand Futures to 2010*.

54. Gibbons and Chandler, *Energy: The Conservation Revolution*.

55. Roger W. Sant et al., *The Least-Cost Energy Strategy, 1978–2000* (Arlington, Va.: Energy Productivity Center, Mellon Institute, 1981).

56. Ross and Williams, *Our Energy*.

57. SERI, *A New Prosperity*.

58. Stobaugh and Yergin, *Energy Future*.

59. National Audubon Society, *The Audubon Energy Plan, 1984* (New York, 1984).

60. SERI is lower than Sant mostly in industry. Part of the difference is in a SERI-proposed "scrap and rebuild" program for our most hopelessly out-of-date and energy-inefficient facilities in financially strapped basic industries such as steel. Also, Sant had industrial output rising more rapidly than GNP, while SERI noted the falling materials-intensity of industrial output and projected a continuation of that trend. SERI also uses higher initial prices than Sant, whose analysis was based on rolled-in energy prices as projected *before* the 1979–80 second oil price leap.

61. The Audubon study does not show efficiency gains as large as does SERI. It has shorter payback times for efficiency investments and no government subsidies. Revised in 1984, it also recognizes time lost under an administration with little interest in energy conservation and renewable energy. The study examines energy use in the year 2000, as do the other studies. It is timely, current, and policy-oriented. It projects rising energy use if various policy actions are not taken, as does the Department of Energy.

62. When we examine space heating in buildings, for example, my estimates are the same as SERI's most efficient buildings, based on 1979 technology, and at energy prices lower than mine. But in the SERI study's terminal year, 2000, "most efficient" is still only a small part of the building stock. I am asking about the 1980 level of energy use in buildings if all of them already were "most efficient."

63. The author's estimate is a rate of 2.9–3.3 percent, of which 2.5–2.8 percent is a true efficiency gain, and the remainder is a result of a continuing relative decline in the most energy intensive sectors of the U.S. economy. The best example is steel. The 1979–84 decline is only slightly overstated by the continuing depression in U.S. basic industry, which is the most energy-intensive sector in the nation. Output in 1984 was still below a "high employment" level by about 5 percent. If one adjusts GNP upward and then estimates the attendant energy use, he can then estimate a 1979–84 "high employment" energy–GNP ratio trend. This calculation does not change the results very much. The 1984 5 percent "shortfall" is calculated by extending the 1979 peak output at 2.5 percent annually. This is the underlying growth rate used by the Department of Commerce, *Survey of Current Business*, December 1983. The 1984 use of energy in

industry is adjusted upward both for "high employment" output and for changes in mix. The latter adjustment is made with factors for efficiency gains and output mix changes developed in a recent industry study. Robert C. Marlay, "Trends in Industrial Use of Energy," *Science*, December 14, 1984. Residential and commercial use is scaled up by a short-term income elasticity of .3. Freight transport rises with industrial output, air transport in proportion to income, and auto transport by a short-term income factor of .3. The largest upward adjustment is in industry—up 13 percent. This is consistent with the disproportionate decline in industrial energy use in the 1984 unadjusted figures.

The resulting energy-GNP ratio for 1984 at "high employment" output is 45,500 btu's per dollar of GNP, as contrasted with 45,000 before the adjustments. Since the 1979 ratio was 59,200, the adjusted rate of efficiency gain is 3.2 percent annually. The unadjusted rate was 3.45 percent.

64. Joel Darmstadter et al., *How Industrial Societies Use Energy* (Baltimore: Johns Hopkins University Press, 1977).

65. D. Yergin and M. Hillenbrand, *Global Insecurity* (Boston: Houghton Mifflin, 1982).

66. Energy Information Administration, *Annual Energy Review, 1983* (Washington, D.C., 1984), p. 41, for nominal prices, adjusted by Consumer Price Index.

67. These are taken from Choe, *A Model of World Energy Markets and OPEC Pricing.* There is a good summary of recent empirical estimates there. See also *Energy: The Next Twenty Years*, p. 90.

68. Actual consumption taken from *Monthly Energy Review*, July 1984. The calculation is a straightforward application of long-run price and income elasticities ($-.8$ and $.85$, respectively) to an output increase of 17.6 percent and a price increase (composite) of 2.5 times 1973 levels.

69. There is a new transformer that uses an amorphous metal core, which reduces energy losses. General Electric estimates the potential savings at 12 billion kilowatt hours, about .5 percent of existing electricity use.

Of much more potential importance are new electricity-generating plant designs. Several have higher efficiencies than those now in use. Fluidized combustion units represent one existing example.

A more efficient possibility is the "Kalina cycle," whose thermodynamic principles are theoretically sound. A prototype is now under construction to test the ideas, and its success would revolutionize the generation of electricity. A good layman's explanation is found in *Popular Science*, August 1986.

70. SERI (*A New Prosperity*, p. 230) estimates that U.S. steel makers use an average of 36 million btu's per delivered ton. This counts electricity at its primary fuel energy value—a point we must always keep in mind when comparing data. When Swedish figures are so adjusted to convert electricity to a primary basis, and to convert metric to U.S. tons, Sweden uses 24 million btu's per ton. That figure is asserted to be susceptible of reduction by half with advanced technology. We settle for a 30 percent reduction (to 16.8 million btu's) already achieved in Japanese plants. See chapter 6 for a discussion of Sweden and Japan. The steel data are found in *Science*, January 1983.

71. So-called "mini-mills" have sprung up around the nation in the last two decades. They use scrap only and melt it with electric arc furnaces. Mini-mills produce a significant

and growing share of American steel and can compete with large mills as effectively as foreign steelmakers. They use about 10 million btu's in making a ton of steel as contrasted with 36 million in large American mills. Continuous casting is one of the technologies that reduces energy consumption in both kinds of mills. It is being adopted more rapidly in the mini-mills. The 10 million btu figure is below the 16.8 million figure given as the most efficient level in note 70. Steel from scrap is inherently less energy-intensive than steel from raw ores. The mini-mill data are reported in the *Wall Street Journal*, January 12, 1981. Mini-mills produce only certain steel products like construction steel. They are gradually moving into other lines as new techniques are developed.

72. Eight million domestically produced autos lightened by 1,000–1,500 lbs. each, mostly by using less steel, yields 4–6 million tons of finished steel. Allowing for the larger amounts of raw steel, and also for weight reductions in light trucks used for personal transportation, brings us to the estimated figure. Domestic steel production has been in the range of 70–100 million tons in recent years.

73. SERI, *A New Prosperity*.

74. Recycled aluminum requires less than 10 percent of the energy used in producing new aluminum from ores.

75. Institute for Energy Studies, Stanford University, *Alternative Energy Futures* (Palo Alto, Calif., 1979), p. 136. This was a listing of reductions that could be achieved in five years.

76. Gulf & Western Corporation, *Quarterly Report to Shareholders*, March 17, 1981.

77. *Governor's Report on Energy*, August 1984, p. 3. (Issued periodically by the Governor's Energy Office, Tallahassee, Florida.)

78. SERI, *A New Prosperity*, p. 231.

79. As an example, Union Carbide announced in 1982 a new process for polyethylene, a major organics product. Energy costs were claimed to be one-sixth to one-ninth of those in conventional processes.

80. Further refinery efficiency gains are discussed but not quantified in Office of Technology Assessment, *U.S. Vulnerability to an Oil Import Curtailment* (Washington, D.C.: U.S. Government Printing Office, 1984). The report points out that refineries can take advantage of conservation techniques for the chemical industries, which it discusses in more detail. Heat recovery and cogeneration are the main possibilities.

81. New developments that were not considered in this analysis or new technologies under development that may contribute to more efficient use of energy in the future include:
—furnace linings that heat up very rapidly, reducing fuel use (reported by McDermott International, *Wall Street Journal*, July 17, 1985).
—specialized applications of electrical process heat in industry. Electricity is generally too expensive to use as an industrial heat source, but some applications are so efficient as to offset the high costs. (A sixfold efficiency gain will offset a threefold cost disadvantage.) Westinghouse advertises induction heating in forging, stamping, and curing metals, and claims energy cost reduction of half or more. Infrared drying of paints is less costly than gas-fired drying chambers.
—in steel, an extension of continuous casting techniques—"thin slab casting"—reduces energy and capital costs. *Fortune*, March 18, 1985.

—a chemical extraction process for solvents, developed by the Department of Energy and an Arthur D. Little subsidiary, has wide application in pharmaceuticals, chemical processing, and waste treatment. Solvents, now recovered from water by energy-intensive distillation, instead can be recovered in a process using condensed or super-critical gases. Possible energy savings reach 80 percent. The process may work for separating ethanol from water, thus reducing the cost (and increasing the net energy yield) of fuel alcohol. *Industry Week*, January 21, 1985, p. 67.

—a cold, superconducting magnet to remove metallic impurities from coal, oil, clays, or other materials. A Pennsylvania firm is developing this device. Electricity is used mainly to keep the magnet cold rather than to maintain the magnetic field. Savings of 80–90 percent are claimed. *Business Week*, June 8, 1985.

3   Renewable Energy in the United States

1. Kendall and Nadis, *Energy Strategies*, p. 19.
2. An excellent treatment, international in scope, is found in Deudney and Flavin, *Renewable Energy—The Power to Choose*. The Union of Concerned Scientists study (Kendall and Nadis, *Energy Strategies*) places more stress on direct solar heat than does this study. The Kendall-Nadis study also has a comprehensive discussion of energy storage.
3. One kilowatt of capacity requires 130–150 square feet in Florida, 90 square feet in California, and 170–200 square feet in New York. A portion of a rooftop would supply an energy-efficient home or low-rise apartment building (or an inefficient one, for that matter, but at needless expense).
4. Bids below $4 per peak watt were reported in a project for the Sacramento Municipal Utility District. The utility rejected all bids, citing costs. Other observers attribute the project deferral to California's oversupply of potential new electricity sources. *Renewable Energy News*, May 1985.
5. Paul D. Maycock and Edward N. Stirewalt, *Photovoltaics* (Andover, Mass.: Brick House, 1981). *Technology Review*, November–December 1984, quoting Battelle Memorial Institute.

   There is a 1985 update of the first work cited: Paul D. Maycock and Edward N. Stirewalt, *A Guide to the Photovoltaic Revolution* (Emmaus, Pa.: Rodale Press, 1985). They report that photovoltaic cell and system prices dropped in real terms over the 1981 period in spite of the cuts in Department of Energy funds. In their earlier work it appeared that photovoltaic system prices would have to fall by a factor of ten in order to be competitive with utility electricity. By early 1985 the required drop in costs was a factor of four. The latter work predicts full competitiveness within a decade (p. 203).
6. Southern California Edison is considering a second 10-megawatt plant with a capital cost of $5,000–$6,000 per kilowatt, a large step down from the first plant. It also is looking toward a larger 100-megawatt plant for 1995, which it believes can be fully competitive. *Renewable Energy News*, June 1985.
7. Luz Engineering, the American subsidiary of an Israeli corporation, has constructed a solar electric plant in California—SEGS-1. The capital cost was about $4,500 per kilowatt of capacity. SEGS-2, now nearing completion, is expected to cost about $3,000 per

kilowatt of capacity. The company believes that subsequent versions will be competitive with conventional plants. *Renewable Energy News*, May 1985.

8. The United States, West Germany, the United Kingdom, the USSR, and Belgium are experimenting with sail-assisted freighters. Japan is furthest along; in 1980 it launched a sail-motor tanker for its coastal trade. The experience with that vessel has aided in the design of larger units. Fuel savings range from 15–50 percent. *Wall Street Journal*, August 4, 1980; *Christian Science Monitor*, April 22, 1983; *Popular Science*, June 1985.

9. *Technology Review*, November–December 1984.

10. Capital cost reductions and improved operating times (which were very low initially) are reviewed by a Pacific Gas and Electric Company consultant and reported in *Renewable Energy News*, July 1985. Presently installed machines, at achievable capacity factors, would generate electricity at 12¢ per kilowatt hour. Machines now being installed would drop that to 11¢.

    A 1980 study by the California Energy Commission estimates that by 1990 wind electricity could be California's cheapest electricity source after hydroelectricity. *Solar Utilization News*, November 1984.

11. The geothermal figures are from Department of Energy, *Inventory of Power Plants, 1984* (Washington, D.C.: U.S. Government Printing Office, 1984). For windpower, more than 1,100 megawatts were in place in California alone by the end of 1985. The December 1984 figure was about 700 kw. American Wind Energy Association, news release, July 16, 1985. *Renewable Energy News*, February 1986.

12. Estimates from annual production, *Solar Age*, various issues; Christopher Flavin, *Renewable Energy at the Crossroads* (Washington, D.C.: Center for Renewable Resources, 1985).

13. In addition to the many vegetable oils long commercialized (sunflower, palm, etc.) there is jojoba, beginning to be commercialized in Arizona. An article in *Science*, September 28, 1984, projects 130,000 tons annually by 1990.

14. Methane yields from feedstocks are climbing as research efforts proceed. Comprehensive reviews may be found in *Science*, e.g., D. L. Klass, "Methane from Anaerobic Fermentation," *Science*, March 9, 1984.

15. The ethanol estimate is from *Chemical and Engineering News*, March 17, 1986.

16. Ethanol must be pure and free of water to stay mixed with gasoline. The distillation process is one of the most energy-intensive steps. Membrane technologies for separation of alcohol and other materials promise considerable reductions in process energy. Some observers see rapid gains in technologies to produce ethanol from wood and other lignocellulosic feedstocks. These could displace grain altogether. Klass, "Methane from Anaerobic Fermentation."

17. Office of Technology Assessment, *Energy from Biological Processes* (Washington, D.C.: U.S. Government Printing Office, 1980). This contains an OTA estimate of biomass energy for the year 2000, giving a range of 6–17 quads. The figures do not include municipal solid waste, estimated by the Council of Environmental Quality at 1 quad, nor from sewage.

18. A pessimistic survey of biomass resources was written by the biomass panel of the Energy Research and Advisory Board (ERAB) in 1982. The casual reader, noting that "biomass can contribute only a small fraction of the energy needed to meet current

U.S. demands. . ." (p. 1), and that potential net energy would be only .8–4.1 quads (p. 2), would conclude that it is hardly worth the bother—a conclusion not uncongenial to the administration being advised by the board.

The report was extensively examined in *Solar Energy*, vol. 30, no. 1, 1983. Since the board accepted the Department of Energy forecast of 100 quads for the year 2000, and estimated the gross biomass resource at 11 quads, the "small fraction" conclusion follows. Eleven quads looks more imposing in a 19–25 quad economy like ours. The panel then properly discusses conversion losses, and reaches its .8–4.1 quad net energy, which seems miniscule against a projected use of 100 quads. Its initial resource figure is lower than OTA's, it makes no allowance for increasing biomass output, its conclusions in comparing gross and net energies are misleading, and it posits conversion efficiencies so low that no one would use them. As the appendix to the previous chapter makes clear, much "industrial" energy use in the United States is energy used to convert fossil fuels, or to mine and transport them. One must be careful to compare only gross with gross, and net with net.

There are many different levels of "gross" and "net." Most studies measure conversion losses in moving from mined or imported mineral fuels into a form suitable for delivery to the end user—crude oil to gasoline or coal into electricity, for example. ERAB goes much further than this, in some cases. Wood for combustion, for example, has a deduction for losses in the furnace or boiler. These are real losses, to be sure, but they render any comparison with conventionally reported energy use very misleading. If these comparisons are to be made, U.S. energy use should be reduced by all conversion losses (25–29 quads out of 70 in 1983) and then again be reduced by combustion losses in boilers and furnaces outside power plants (these already are deducted above).

As in all matters, one must read the fine print before accepting conclusions.

19. *Technology Review*, April 1983.
20. Deudney and Flavin, *Renewable Energy—The Power to Choose*, p. 85.
21. Kenneth C. Brown, "Reexamining the Prospects for Solar Industrial Process Heat," in Jack M. Hollander and Harvey Brooks, eds., *Annual Review of Energy, 1983* (Palo Alto, Calif.: Annual Reviews, Inc., 1983).
22. Keynote address, "Solar Energy: Potential, Problems and Implications," by H. M. Hubbard, director of SERI, to the American Society of Mechanical Engineers, Solar Engineering Division. Published in *Journal of Solar Energy Engineering*, August 1984.
23. As usual, technology does not stand still. In reviewing renewable electricity technologies, I had ignored wave power. The U.S., U.K., and Japanese research programs suggested that the source was far from commercialization. Now Norway has two operating prototypes, claims that electricity from its installations would cost a very competitive 4–5¢ per kilowatt hour, and is seeking customers. *Wall Street Journal*, July 19, 1985; *New Scientist*, July 11, 1985. Ironically, one Norwegian plant is from a British design, and Britain is abandoning wave-energy research.
24. Energy use in 1983 was reported at 70.5 quads; 1984 use at 73.7 quads. When energy sources not officially reported are added, these figures come to about 73 and 76.7 quads, respectively. Biomass use is estimated at 2.6 quads by the Department of Energy and 2.7 quads by Klass (Donald L. Klass, Institute of Gas Technology, "1984 Update," paper presented at Ninth Annual Conference, Energy from Biomass and

Waste, Lake Buena Vista, Florida, March 1985). Flavin, *Renewable Energy at the Crossroads*, uses a 3-quad estimate. In addition to wood, the biomass estimates include fuel alcohol, methane from wastes, and garbage burning for electricity.

25. The solar resource is so large relative to these demands that it may be considered limitless. The constraint is the cost of equipment for the collection and conversion of energy. Photovoltaic electricity, with presently available equipment for converting direct current to alternating current, can supply up to 25 percent of a grid's power without degrading the quality of the current. (Florida Solar Energy Center, *Annual Report 1984*.) This limit, and that imposed by the time of availability as modified by storage, are the only practical technical limits on photovoltaic electricity; the resource base is enormous.

    The gross biomass resource has been estimated at 18 quads or more (OTA, *Energy from Biological Processes*). The Gas Research Institute projects biomass-based methane alone to 10 quads by 2000. *Methane from Biomass and Waste Research* (Chicago, 1984). This much methane is not likely to be demanded in an energy-efficient economy, and some of the feedstock will be needed for methanol.

    Geothermal potentials are discussed in National Research Council, *Geothermal Resources and Technology in the United States*, Study of Nuclear and Alternative Energy Systems, Supporting Paper 4 (Washington, D.C.: National Academy of Sciences, 1979). The wind-electric potential has been most thoroughly assessed in California, where some 13,000 megawatts could be in place by 2000 if that much electricity from one source were to be needed. Site-measurement work like that done in California is badly needed in the rest of the nation. The resource is known to be large in the Northeast, the Great Plains, and on the Texas coast.

26. The Department of Energy estimates that methanol from wood now would cost $13 per million btu's, as compared with gasoline at about $6–$7 wholesale. Gasoline would be more like $10 at long-run marginal supply prices. The department's target for wood-based methanol is $8.60 per million btu's—a competitive figure against gasoline at $10. Department of Energy Biofuels Program Goals, reported in *Renewable Energy News*, February 1985.

    Progress is slow because fossil-fuel-based methanol is so cheap. There is a worldwide excess capacity for methanol.

27. A useful survey of some dozen projections is found in John P. Holdren, "Renewables in the U.S. Energy Future: How Much, How Fast?" *Proceedings, National Conference on Renewable Energy Technologies* (Honolulu: University of Hawaii, 1980). Even President Reagan's Department of Energy is projecting 10 quads of renewable energy in 2000.

28. CONAES study (*Energy in Transition 1985–2010*). Figures are from supply panels for renewable sources, hydro and geothermal, which were treated separately.

29. The appendix to chapter 2, which summarizes efficiency gains, also estimates remaining electricity demand as a share of reduced total energy demand. Whereas electricity is strictly required for only about 8 percent of present final demand, it represents a larger share of end-use demand in virtually all studies of energy-efficient futures. This is partly because space-heat needs are so much reduced and because such large efficien-

cy gains are available in autos. Efficiency gains in electric-specific applications also are quite large, but not as large as these.

30. The forest products industry is now a bit over 50 percent self-sufficient in energy. This can be increased beyond 100 percent—energy can be "exported" to other industries—via further efficiency gains in the forest products industry itself and by the use of still more materials now wasted.

31. "Biogas" is produced from animal wastes and sewage at present. Nearly any plant or animal waste is a potential feedstock. Digesters produce a mixture of methane (natural gas), carbon dioxide, and other gases such as hydrogen sulfide. For many uses and for pipeline transmission, the gases other than methane must be filtered out. Brooklyn Union Gas Company now is doing this commercially with biogas from garbage dumps in the New York area.

32. Fred Sanderson, "The High Cost of Gasohol," *Resources*, July 1981.

33. One such process has been developed at the University of Florida. Professor Wayne Smith, personal communication.

34. For a discussion of oil-bearing plants, see note 13 of this chapter.

35. SERI, *A New Prosperity*, pp. 318–19.

36. Donald F. Othmer of the Polytechnic Institute of New York in Brooklyn asserts that methanol is 61 percent more efficient per btu than is gasoline, in properly modified engines. Some 1.24 gallons of methanol would then provide the vehicle fuel equivalent of one gallon of gasoline, not two gallons as implied by relative energy contents. These higher equivalencies are not used in the foregoing calculations. Othmer cites research under way on new engine technologies that would use still more of the energy in methanol. *Alcohol Week*, May 6, 1985.

37. Energy Research Advisory Board, report, cited and criticized in chap. 3, n. 18 above.

38. Quoted in SERI, *A New Prosperity*, p. 253.

39. Hubbard, address, cited in chap. 3, n. 22 above.

40. Petr Beckmann, "Solar Energy and Other 'Alternative' Energy Sources," in Julian L. Simon and Herman Kahn, eds., *The Resourceful Earth* (Oxford: Basil Blackwell, 1984).

41. *Public Utilities Fortnightly*, September 28, 1978.

42. Ethanol plants of recent vintage are more energy-efficient. They have favorable net energy outputs (measuring fossil fuels used in processing and used indirectly in growing grain). Flavin, *Renewable Energy at the Crossroads*.

43. With the hydroelectric contribution so large, and with additional pumped storage capacity, it is hard to imagine demand profiles that could not be accommodated. Some modeling of utility load profiles and wind-photovoltaic performance is badly needed in each region of the nation. This would determine the extent to which wind-photovoltaic contributions could be increased over my rough estimates and the amount of additional pumped storage that might be needed.

44. M. Diesendorf, an energy analyst with the Australian Commonwealth Scientific and Industrial Research Organization, reported on this work with various colleagues in *Soft Energy Notes*, February–March 1981.

45. Large-scale compressed-air storage in underground chambers has been discussed,

mostly as a concept. It is covered in the Union of Concerned Scientists study (Kendall and Nadis, *Energy Strategies*, pp. 217–19). One such installation exists in West Germany and another is nearing completion in Italy. IEEE, *Spectrum*, January 1985, p. 73. This technology may be appealing in areas lacking adequate hydroelectric capacity or pumped storage.

46. This is a point not yet grasped by many critics of renewable energy futures. It is simply assumed that growth and prosperity require large and growing quantities of energy so that renewable energy advocates are automatically seen as antigrowth.

47. National Research Council, Committee on Nuclear and Alternative Energy Systems, *U.S. Energy Supply Prospects to 2010* (Washington, D.C.: National Academy of Sciences, 1979), p. 181.

48. OTA, *Energy from Biological Processes*. Other possible additions or substitutions include energy crops, enhanced yields, and aquaculture, especially in warm climates.

4 State, Regional, and Local Issues

1. Beckmann, "Solar Energy and Other 'Alternative' Energy Sources."

2. Among the innovative plans for the use of existing dams without disrupting other uses is a barge-mounted generator in Wisconsin. It operates in a shipping lock on the Mississippi River and is moved aside when vessels use the lock. One installation will yield 10 megawatts; 50 dams, 1,200 megawatts. *Alternate Sources of Energy*, September–October 1984.

3. OTA, *Energy from Biological Processes*, gives an estimated gross biomass resource for the United States of 17 quads. This was noted in chapter 3. Of this, about 2 quads is most readily suited to methane production.

4. SERI, *A New Prosperity*, p. 269. Figures there are for average days, in kilojoules per square meter. I have converted to more familiar btu's per square foot.

5. California Energy Commission, *Securing California's Energy Future*, p. 53.

6. As this manuscript was nearing completion, the district rejected all bids on the third phase, saying that they were too high. Sources close to the bidders question this statement. It appears that the utility is having second thoughts about completing the plant. This is consistent with the general rethinking in California of electricity supply sources. So much potential supply from conservation, cogeneration, and renewable sources is available that priorities are being reexamined.

7. California Energy Commission, *Securing California's Energy Future*, p. 54.

8. Northwest Power Planning Council, *Regional Conservation and Electric Power Plan, 1983* (Portland, Oreg.: Northwest Power Planning Council, 1983). A much more comprehensive study for Oregon is found in Oregon Alternate Energy Development Commission, *Future Renewable, Final Report* (Salem: State of Oregon, 1980). This report deals with the demand for heat energy as well as the demand for electricity and concludes that, of new energy demands, more than 100 percent could be met from renewable sources.

9. The Oregon Alternate Energy Development Commission report, pp. 3–9, gives a cost-ranking of various supply options. Five thousand megawatts of saved energy is the least expensive.

10. Ibid.

11. R. Katz et al., *Energy in New England: Transition to the '80's* (Washington, D.C.: New England Congressional Institute, 1981).

12. New York City Energy Office, *Energy Consumption in New York City* (New York: New York City, 1981).

13. Beckmann, "Solar Energy and Other 'Alternative' Energy Sources," p. 418.

14. We can translate these energy data into terms used in this study. One watt per meter times 8,760 hours per year gives 8.76 kilowatt hours. At 3,413 btu's per kilowatt hour, this is about 30,000 btu's. A density of energy use of 7.5 w/m², the West German urban area figure, is thus about 225,000 btu's per square meter per year, or about 21,000 btu's per square foot per year. Since average insolation (receipt of solar radiation) is about thirteen times this figure in Germany, Beckmann rests his case on low conversion efficiencies.

15. New York City Energy Office, *Energy Consumption in New York City.*

16. Ibid., p. 20.

17. Marine fuel efficiency also can be improved through increased engine efficiency and design features of hulls and propellers. Some highly sophisticated control systems for modern sails also have provided some substitute power for fuel oil. Marine fuel accounts for about 6 percent of the energy sold in New York.

18. The Environmental Defense Fund has raised questions about the garbage-to-electricity program on several counts. One has to do with air pollution, and specifically the threat of dioxin. There are, apparently, technologies to overcome these concerns. Second, the fund and others point out, as does this study, that recycled paper has a higher raw material than fuel value. Accordingly, the energy potential is reduced by 25 percent, from New York's 72 trillion btu's to 54 trillion. *EDF Letter*, March 1985. This, of course, saves still more energy elsewhere. About 2 trillion btu's can be added back with more extensive urban tree-planting.

19. Brooklyn Union and Texaco have one operating methane extraction facility and are starting another. Other landfills under consideration could bring total production to 5 percent of Brooklyn Union's present sales. *Wall Street Journal*, June 28, 1984.

20. *New York Times*, October 25, 1982.

21. New York City Energy Office, *Energy Consumption in New York City*, p. 33. Calculated from area given for floor space, adjusted for average number of floors per building.

22. The 13–18 trillion btu figure is based on 110–115 million square feet, annual insolation of 400,000 btu's per square foot, and a conversion efficiency of 10 percent to electricity. Electricity is then stated in primary energy equivalents. The calculation is checked by taking square-foot and power outputs for other locations and adjusting for New York City sunlight conditions.

23. New York already gets nearly this much electricity from existing hydro facilities.

5   What of Coal, Synfuels, and Nuclear Power?

1. IEA, *World Energy Outlook* (1982). Figures are given in "metric tons of oil equivalent." They are converted here to quads.

2. Workshop on Alternative Energy Strategies, *Energy: Global Prospects 1985–2000* (New York: McGraw-Hill, 1977).

3. ERDA, *A National Plan.*

4. The required buildup of the various components of the nuclear system is discussed in WAES, *Energy: Global Prospects 1985–2000*, and in virtually all of the energy analyses of the 1970s.

5. The importance of viewing nuclear reactors as only one component of an integrated system is stressed in Irvin C. Bupp and Jean-Claude Derian, *The Failed Promise of Nuclear Power* (New York: Basic Books, 1981).

6. The U.S. "gap" is calculated from data in the *Survey of Current Business*, December 1983. The European gap is estimated from extrapolations of 1973–80 growth compared with an estimate for 1985 as it now seems likely to turn out.

7. The U.S. calculation is given in chap. 2, n. 63. A 5 percent upward adjustment in GNP leads to a 6.5 percent upward adjustment in energy use, since the energy-intensive industrial sector is so depressed. Europe is about 10 percent below capacity output; its energy consumption estimate for a "high employment" output is scaled up 13 percent. Western Europe,with a larger population than the United States, uses about half as much energy.

8. The adjustment described in the preceding note 7 dealt with total energy. Oil provides about 40 percent of that total. At the margin, it is perhaps 50–60 percent. An additional 5–6 quads of oil to accommodate "high-employment" outputs in the OECD nations, is about 2.5–3 million barrels a day. Surplus capacity in the world oil industry is more like 15 million barrels a day.

9. U.S. coal use is from *Monthly Energy Review*, March 1985. OECD use is from OECD, *Energy Balances of OECD Countries, 1982–1983* (Paris, 1985).

10. About 250 reactors have been announced or ordered in the United States. (The tabulation in the *Monthly Energy Review* gives a peak number of 236 in 1975, but some of these were canceled before others were ordered.) In mid-1986, about 100 units are operating or are in start-up condition. An additional seven have operated but have ended their useful lives. Of the remainder, 109 have been canceled and 35 are still listed as under construction. Of these, another 15 or so may be canceled. The eight reactors that have ended their useful lives all did so prematurely, except for the first commercial reactor at Shippingport, Pennsylvania. It is now being decommissioned.

11. Synthetic fuels, incidentally, were not expected to be very significant until around the year 2000, but some production was expected by 1985. There is virtually none. The technology has proved to be more difficult and less readily commercialized than energy experts thought in the 1970s.

12. Reprocessing plants operate in the United Kingdom, France, and Japan. The present U.K. facility was not designed for waste from reactors now being built. The one U.S. facility that has operated was shut down in 1972. Two other U.S. facilities have not operated. The foreign facilities have a design capacity of 610 tons of spent reactor fuel. The technology is very demanding. Plants run at 10–35 percent of their design capacity, and they do not last very long. No operating costs have ever been published—only estimates are available. Operating rates and plant lives are from a study by Arjun Makhijani, a nuclear engineer. The study was summarized in the *Washington Post*, July

10, 1985. The relatively low operating and high unit cost rate for the largest, the French unit at La Hague, is also reported by D. Collingridge, "Lessons of Nuclear Power: French 'Success' and the Breeder," *Energy Policy*, June 1984.

If reprocessing plants operate at 10 to 30 percent of design capacity, then the world's effective reprocessing capacity is 61–180 tons per year. This will accommodate 2–5 percent of the spent fuel being produced. The rest of the world is in a situation not very different from that of the United States. Most (in the United States, all) of spent reactor fuel is going into temporary storage.

13. The West Valley, N.Y., plant was closed in 1972 for major alterations. It had worked poorly, averaging 10 percent of design capacity. Costs of renovation to improve safety and reliability were so high that the plant was abandoned. The state of New York and the federal government are sharing the cost of removing radioactive material. General Electric's plant at Morris, Illinois, did not work. Rather than redesign and rebuild virtually the entire plant, GE abandoned it. The plant at Barnwell, S.C., has not been completed. Private industry has been unwilling to assume the considerable risks. The Reagan administration, committed both to nuclear power and to free markets, has in this instance come down on the side of free markets.

14. WAES, *Energy: Global Prospects 1985–2000*, p. 209.

15. Reprocessing was abandoned in the United States since it was more costly than expected, and uranium was more abundant than expected. The costs exceeded the benefits. In the United States the threat of nuclear proliferation that accompanied reprocessing also was a major factor in its negative appraisal. The authors of a Ford Foundation-sponsored study pointed out in 1979 that reprocessing costs were understated by a factor of ten (*Energy: The Next Twenty Years*). That estimate is consistent with a tenfold escalation in the inflation-adjusted costs of French reprocessing—or, rather, the tenfold increase in published estimates. If it is not economical in the United States, it can hardly be so anywhere else, since foreign experience is similar to that of the closed U.S. plant.

16. The German pilot breeder was started in 1973 for completion in 1978 with an expected cost of $550 million. By 1982 the expected completion date had moved to 1986 and the estimated cost to $2 billion (*Christian Science Monitor*, March 4, 1982). The Japanese pilot breeder, still in planning, has seen its estimated cost go from $1.6 billion to $2.5 billion. As in Germany, cost-sharing between the national government and the utilities is at issue (*Nature*, October 28, 1984). Britain continues to operate a pilot breeder but has delayed for several years a decision to proceed with a larger one.

17. France began operating an experimental breeder in 1967, a 250-megawatt pilot breeder, "Phénix," in 1973, and is now completing a large 1,200-megawatt demonstration breeder, "super-phénix." The cost exceeds $2 billion in 1985's distorted exchange rates. The cost would be $2.5–$2.7 billion at exchange rates more likely to prevail over time.

18. The U.S. government is decommissioning the nation's first commercial reactor—the small (60 MW) Shippingport plant. The estimated cost is $66 million. When that task is completed, estimates for larger, long-lived reactors will be on much firmer ground than they are now. They still will be estimates until a large (1,000 MW) reactor is actually dismantled.

19. Ronnie D. Lipschutz, author of a report for the Union of Concerned Scientists, writes that "the theoretical basis for safe radioactive waste management does exist, although . . . it is by no means a complete basis" (*Radioactive Waste: Politics, Technology and Risk* [Cambridge, Mass.: Ballinger, 1980], p. 111). In reviewing the history of waste management in the United States he is less positive.

20. Data for Europe and Japan are from the World Bank, *World Development Report, 1985* (Washington, D.C., 1985).

21. *Economist*, July 6, 1985, quotes from a scholarly but unpublished paper drawing on research at the Polish Academy of Science. The USSR uses three times as much energy per unit of GNP as does France and twice as much as does the United Kingdom. This is understated, since official ruble exchange rates are used in comparing national outputs. Hungary, the most energy-efficient East bloc nation, uses twice the energy per unit of national product as does France.

22. *Monthly Energy Review*, March 1985.

23. Nuclear plant costs in the United Kingdom and West Germany are similar to those in the United States and have undergone similar rates of escalation in real terms. French costs are lower but are rising at about the same rates. The French case is discussed in the appendix to this chapter, and sources are given there.

24. OECD, *Energy Balances of OECD Countries, 1973–1975* (Paris, 1976).

25. The Duke Power Company manages in this fashion. Ninety-seven percent of its electricity is from coal and nuclear plants. Its hydroelectric facilities contribute another 3 percent and are used almost exclusively for peak demands (*Annual Report*, 1984). Power interchanges with other utilities also help.

26. Department of Energy-sponsored research is aimed at decreasing battery weight per kilowatt of storage, lowering costs, raising efficiency, and lengthening life. No battery suitable for widespread use in vehicles has yet resulted. Gulf & Western has been working for some years on a zinc-chloride battery. It will be installing 2 megawatts for the isolated West Berlin utility. For stationary applications, weight reduction is less important, but efficiency, cost, and longevity have thresholds for commercialization. *Economist*, August 24, 1985.

27. The only synthetic fuel facility now working in OECD nations is one that produces oil from tar sands in Canada. Costs were reported at $35 per barrel in 1982 (*World Energy Outlook*, 1982). Raising the operating time would reduce this figure. Unocal has a plant to produce oil from shale at a rate of 10,000 barrels per day. The federal price guarantee is $42.50 per barrel. The plant has not yet operated successfully and is not likely to be able to produce at that cost (*Investor's Daily*, August 5, 1985). Higher estimates are found in IEA, *World Energy Outlook* (1982), p. 418.

28. Decommissioning costs were estimated at $400 million for large reactors by industry sources in 1980 (*Public Utilities Fort-nightly*, September 25, 1980). That would be at least $500 million in 1986. These are not negligible parts of the cost of electricity as some earlier studies assumed. One of the Ford-sponsored reports (*Energy: The Next Twenty Years*) shows decommissioning costs per kilowatt hour of .002¢. When one uses shorter reactor lives, lower operating rates, and higher decommissioning costs, this estimate can be off by a factor of 100 or more. Decommissioning could add .1 to .5¢ per kilowatt hour. That looks like a small figure, but investment decisions have been made on smaller differences.

29. Charles Imbrecht, chairman, California Energy Commission, address at Twelfth Energy Technology Conference, Washington, D.C., reported in *Renewable Energy News*, May 1985.
30. Klass, "1984 Update."
31. Enrichment is done by a European consortium, Eurodif. The plant is located in France; that nation is the largest customer.
32. I have relied on a discussion of the Ninth Plan, and alternatives, in Confederation Française Democratique du Travail (CFDT), *Group Conféderale Energie Le Dossier de l'énergie* (Paris: Editions du Seuil, 1984).
33. Lovins, *Soft Energy Paths.*
34. *Nature*, May 26, 1983, p. 273. By late 1984 the projected overcapacity in 1990 was equal to the output of eight plants. *Wall Street Journal*, November 18, 1984.
35. Jim Harding, "The French Nuclear Debacle," *The Ecologist*, vol. 14, no. 3, 1984.
36. Ibid., p. 106. Calculated from Electricité de France data.
37. Ibid.
38. CDFT, *Group Conféderale*, pp. 51 and 53.
39. Harding, "The French Nuclear Debacle," p. 109.
40. In recent years construction outlays have run about 40 billion francs, and fixed, nonresidential investment at about 400 billion.
41. The slogan conveying the all-or-nothing approach openly espoused by French energy planners is "toute éléctrique, toute nucléaire."
42. Irvin C. Bupp and Jean-Claude Derian, *Light Water: How the Nuclear Dream Dissolved* (New York: Basic Books, 1978).
43. Bupp and Derian, *The Failed Promise of Nuclear Power*. This is a paperback, revised edition of their work cited in the preceding note.

6   Other Industrial Nations

1. Darmstadter et al., *How Industrial Societies Use Energy.*
2. Thomas B. Johansson et al., "Sweden Beyond Oil: The Efficient Use of Energy," *Science*, January 28, 1983.
3. Amory Lovins et al., *Least-Cost Energy: Solving the CO_2 Problem*. (Andover, Mass.: Brick House, 1981). While this is aimed at worldwide issues and fossil fuel dependence, the core example is a case study of the Federal Republic of Germany, for which Florentin Krause, one of the coauthors, is largely responsible.
4. David Olivier and Hugh Miall, *Energy Efficient Futures: Opening the Solar Option* (London: Earth Resources Research Ltd., 1983).
5. Studies for individual Canadian provinces are carried in several issues of *Alternatives*, 1979 and 1980.
6. Groupe de Bellevue, *Project Atler* (Paris, 1978).
7. World Bank, *World Development Report, 1985* (Washington, D.C., 1985), and Commission of the European Communities, *A European Energy Strategy* (Brussels: EEC, 1984), European Investment Bank, *Information*, May 1985. The latter reference describes a new garbage-fired district heating system in Copenhagen that will, alone, replace oil for 3 percent of the nation's total energy use. It then proceeds to discuss energy in conventional terms with no mention of renewable sources.

8. European hydroelectric capacity grew from 99,400 MW to 140,600 MW over the 1970–81 period (UN Statistical Yearbook, 1981). French hydroelectric production, for example, rose more than 60 percent from 1973 to 1979, yet its 1990 energy supply projection calls for *less* hydroelectricity in 1990 than in 1979. Ministry of Industry, *Le Bilan Energétique a l'Horizon 1990* (Paris, 1980).

9. The World Bank, always cautious, uses an estimate of 5–10 percent. U.K., French, and U.S. experience suggest that the higher range is appropriate. World Bank, *The Energy Transition in Developing Countries* (Washington, D.C., 1983).

10. The 30,000-MW figure is derived later in this chapter in more detailed examinations of per capita usage with high energy efficiency in Sweden and the United Kingdom. Since we assume no long-run disadvantages in per capita income for any nation, electricity usage would vary mainly with relative price. Yet intra-European trade in electricity is rising rapidly (and will take place shortly between France and the United Kingdom via under-the-channel cable). Large price differentials are not likely to persist.

11. The importance of resource assessment is underscored by the experience of France and Spain with respect to windpower resources. Energy planners there either ignored windpower altogether or assumed negligible possibilities. Even renewable energy enthusiasts concentrated on the nations mentioned in the text. Yet French wind-electric potential goes well beyond the known high-potential coastal areas of Normandy and Brittany. Areas extending quite far inland from the Mediterranean also were found to have a high potential. M. Luneau, *Les Energies Nouvelles: Qu'en Espérer?* (Paris: La Documentation Française, 1982), p. 54. Spanish investigators discovered, to their surprise, a potential of some 9,000 MW. *El Vent: Energia eòlica a Catalunya* (Barcelona: Generalitat de Catalunya, 1984).

12. *Economic Handbook of the World*, 1981.

13. Johansson et al., "Sweden Beyond Oil."

14. Olivier and Miall, *Energy Efficient Futures*.

15. Darmstadter et al., *How Industrial Societies Use Energy*.

16. Watt Committee Report, quoted in *Renewable Energy News*, August 1985.

17. The recently completed 1,681-megawatt pumped storage unit in Wales is discussed in *New Scientist*, May 19, 1983. Pumped storage is the key to Britain's renewable electricity future, since the country's hydroelectric capacity is relatively small. Tidal, wind, and photovoltaic power can benefit from a large pumped storage system, unless still cheaper technologies emerge. Dinorwig, the 1,681-megawatt unit mentioned above, cost about £250 per kilowatt of capacity ($400–$500).

18. *Financial Times*, April 2, 1983, observes that government activity with respect to the tidal facility might be construed as a weakening of the commitment to nuclear power.

19. *Electronics and Power*, 1982. This source estimated that 20 percent of the nation's electricity could come from wind without assimilation problems in the existing grid. That was before the first large pumped storage unit was added.

20. One quad of electricity (primary energy equivalent) may be thought of as 11,000 megawatts of capacity running all the time (100 percent capacity factor). If the technologies chosen for the U.K. renewable electricity supply averaged 40 percent, then 41,000 megawatts would be required. With 2,800 (hydro) and 7,200 (tidal) available, and with 6,000–9,000 from cogeneration (mostly biomass), the remainder for wind and

photovoltaic begins to appear manageable with the additional pumped storage facilities that can be built.

21. The Olivier estimates are supplemented by the work of other researchers. M. R. Jones estimates about .5 quad from animal wastes, crop wastes, farm woodlands, and the use of now derelict lands (S. P. Carruthers and M. R. Jones, "Agriculture's Potential Contribution to U.K. Fuel Supplies," in A. Straub et al., eds., *Energy from Biomass*, Second EEC Conference (London: Applied Science Publishers, 1982). When one adds the contributions of the forest products industry, urban wastes, and energy farms, their estimates are consistent with Olivier's.

22. U.S. data from OTA, *Energy from Biological Processes;* ERAB (see chap. 3, n. 18).

23. At British insolation levels, this implies some 800–900 million square meters, a figure not greatly different from rooftop area. Not all collectors will go on roofs, nor will all roofs take collectors. This is mentioned solely to put into perspective a space need that may seem too large in a densely populated nation.

24. Japan's population is approximately twice that of the United Kingdom.

25. This same assumption is made by Juji Watanuki, "Japanese Society and the Limits of Growth," in Yergin and Hillenbrand, eds., *Global Insecurity*, p. 179.

26. World Energy Conference, *Survey of Energy Resources* (New York: U.S. National Committee of the World Energy Conference, 1974).

27. Wilfred L. Kohl, ed., *After the Second Oil Crisis: Energy Policies in Europe, America and Japan* (Lexington, Mass.: D. C. Heath, 1982).

28. Study by Mitsubishi Research Institute, quoted in H. Tsuchiya, "Soft Energy Planning for Japan," *Soft Energy Notes*, February–March 1981.

29. Medard Gabel, *Energy, Earth and Everyone* (New York: Audubon Books, 1980). Government projections of 1980 call for 3,500 megawatts by 1990. Kohl, *After the Second Oil Crisis.*

30. Tsuchiya, "Soft Energy Planning for Japan." One Japanese investigator, studying methane potentials, examined yields from sweet potatoes. These give quite high returns per unit of land. Even so, a quad of methane would require 2,000,000 hectares, about 37 percent of Japan's arable land. Some energy crops might be tolerated, but Japan already imports about half its food. More likely is the introduction of higher-yielding trees in forest areas. The sweet potato study is given in Straub et al., eds., *Energy from Biomass.*

31. Laurent Piermont, *L'énergie Verte* (Paris: Editions du Seuil, 1982). Piermont does not tabulate all of his findings, but he identifies some 40 million tons of petroleum equivalent (1.6 quads) as a potential. This figure is based mainly on existing forests and waste streams and requires relatively little land devoted to energy crops to reach the 2-quad figure.

32. Japan maintains self-sufficiency in rice (though not in total food requirements) by maintaining a domestic price for rice much higher than the world price. Europe has attained overall self-sufficiency in food (as well as surpluses in some foodstuffs) with food prices some 7 percent higher than U.S. prices. Data from Irving B. Kravis et al., *World Product and Income: International Comparisons of Real Gross Product* (Baltimore: Johns Hopkins University Press, 1982). This study used 1975 data to establish purchasing power equivalents at which time the United Kingdom was still adjusting to the

relatively higher continental price supports. These are retail prices for selected but numerous foods, as part of a general purchasing power comparison.

Among renewable resources for energy, only solar heat in Northern Europe (lower capacity factors on collectors) and biomass fuels (denser population) should be higher in cost than in the United States.

7   Third World Energy Futures

1. See, for example, Yergin and Hillenbrand, eds., *Global Insecurity*.

2. An important study recently published examines an energy-efficient development path in industry, transport, households, and commerce. Developing countries could enjoy a standard of living, and level of energy services, comparable to those of Western Europe in 1980, but with only 10 percent more energy than they now consume. Energy demands at that low level are determined by examining each end use and looking at the most energy-efficient way to support that end use. The cost-effectiveness criterion is a little looser than mine—technologies that are now cheaper than new supply or those that have good prospects of becoming so within the decade are considered. Jose Goldemberg et al., *An End-Use Oriented Global Energy Strategy*, Report #179 (Princeton, N.J.: Princeton Center for Energy and Environmental Studies, 1985).

3. World Energy Conference, *Survey of Energy Resources*. Another excellent source for all energy forms is Deudney and Flavin, *Renewable Energy—The Power to Choose*.

4. For 1970 and 1981, *U.N. Statistical Yearbook, 1982*. For other data, see Deudney and Flavin, *Renewable Energy—The Power to Choose*.

5. For 1965 and 1981, *U.N. Statistical Yearbook, 1982*. The estimate of potential is from J. Dunkerly et al., *Energy Strategies for Developing Nations* (Baltimore: Johns Hopkins University Press for Resources for the Future, 1981).

6. Geothermal Resources Council, *Bulletin*, June 1984. World-wide geothermal installed capacity had reached 3,770 megawatts, with a projected capacity of 6,000 megawatts by 1986. The World Bank projects above 4,000 megawatts in the early 1990s for the developing nations. This figure may be reached sooner if plans are not delayed too long by falling petroleum prices and debt problems in developing nations.

7. Christopher Flavin, *Wind Power: A Turning Point*, Worldwatch Paper 45 (Washington, D.C.: Worldwatch Institute, 1981).

8. *Technology Review*, November–December 1982.

9. Dunkerly et al., *Energy Strategies for Developing Nations*.

10. Modern charcoal kilns are twice as efficient as traditional ones, converting 40 percent of the wood's energy content to charcoal as contrasted with 20 percent or less in traditional kilns. Capturing gases and oils from the wood can raise conversion efficiencies to 80 percent. World Bank, *Renewable Energy Resources in the Developing Countries* (Washington, D.C., 1980), p. 11.

11. Successful reforestation projects in Korea (the outstanding case), China, and states in India are described in Deudney and Flavin, *Renewable Energy—The Power to Choose*.

12. Centrally planned economies in general seem to be considerably more energy-intensive than the rest of the world. China used, in 1979, 2.5 times as much energy per unit of output as did other developing nations, none of which use energy at levels of efficiency

targeted earlier in this work. For more extensive comments on China, see Robert P. Taylor, "Energy Conservation," *China Business Review*, January–February 1982. See also H. Kohl and E. Segura, "Industrial Energy Conservation in Developing Countries," *Finance and Development*, December 1983.

13. The authors of the Princeton study referred to in note 2 of this chapter use a lower estimate—30,000,000 btu's per capita per year. This, however, is end-use energy—after conversion and distribution losses. The 50,000,000 btu's per capita that I use is about 37,000,000 btu's per capita after these losses. My allowance is therefore about 20 percent higher than that in the Princeton study.

14. Automobile ownership seems to saturate at about two persons per vehicle. The figure in Western Europe is about three per vehicle, with population densities influencing the figure as well as income. Italy, less densely populated than the United Kingdom, has more autos per capita despite its lower income. Japan has fewer still, as would China when habitable land is considered.

15. The lower estimate is from Dunkerly et al., *Energy Strategies for Developing Nations*. The higher uses a larger share of the theoretical maximum.

16. Ibid. These estimates consider forests, crop residues, and animal manures. Municipal solid wastes, other processing wastes, and energy crops are not included. Only existing forests are considered.

17. Energy crops would not have been suggested for China five years ago, since food is the first priority. The increase by half in grain output in six years and the attainment of food self-sufficiency open up new prospects. Grain-based ethanol with by-products suitable for animal feed suggest ethanol potential of .3–.5 quads. Sugarcane already is grown; varieties bred for energy and sugar yield almost as much sugar per acre and much more energy feedstock. This offers another source of energy without competing with food acreage.

   Urban wastes are, understandably, not considered in the Resources for the Future study. We are here considering a high-income China with living standards similar to those of Japan or Western Europe in 1980. Municipal solid waste in Copenhagen provides a gross output of 25 trillion btu's or 16,000,000 btu's per capita. New York figures show 10,000,000 btu's per capita. Allowing for further recycling of paper might reduce this to 8,000,000. In China, only 20 percent urban, this provides 1.6 quads. A higher-income China would probably be more urbanized, so that a figure of 2 to 4 quads would be more appropriate. Urban sewage systems, if designed at the outset to do so, can provide about 1,000,000 btu's per capita, or .2–.4 quads in China. Among the design requirements are the pretreatment or separation of industrial wastes that might interfere with the methane digesters.

18. Dunkerly et al., *Energy Strategies for Developing Nations;* World Bank, *The Energy Transition in Developing Countries*.

19. Roger H. Charlier, *Tidal Energy* (New York: Van Nostrand, 1982).

20. United Nations, *Economic and Social Survey of Asia and the Pacific, 1980* (New York, 1981), p. 77.

21. Preliminary U.S. data suggest a land requirement of about 3,000 acres per million people to reach the same standards of water purification as is achieved by modern sewage plants with tertiary treatment. The water plants effectively double the energy

yield of sewage. If any land is to be devoted to energy crops, it may as well be devoted to some of the highest-yielding plant life on earth. The same area also purifies water for reuse and recaptures nutrients for fertilizer.

Such facilities for 400,000,000 people would require 12 million acres (4.8 million hectares) and provide .8 quads. Treatment ponds can be put almost anywhere—good cropland need not be taken.

Land requirements are estimated from *Environment*, June 1980.

22. World Bank, *World Development Report, 1985*.
23. World Energy Conference, *Survey of Energy Resources*; World Bank, *The Energy Transition in Developing Countries*.
24. Stream flow, and therefore hydroelectric output, varies a great deal over the year, with the lowest flow in the late summer. Photovoltaic and solar pond outputs are highest at that time, giving some complementarity to the system and permitting the development of a larger fraction of the year-round hydro potential.
25. Solar ponds are quite land-intensive. Israeli and California experience suggest a land requirement of 250 square kilometers for 1,000 megawatts of capacity. *Technology Review*, November–December 1982. Senegal has 196,000 square km; siting, not total land area, is the constraint.
26. *The Development of the Gambia River Basin*, United Nations Development Program, Multi-Disciplinary, Multi-Donor Mission, Final Mission Report (New York, 1980).
27. One of the principal crops is peanuts. Hulls from processing could provide 7–8 trillion btu's—1 percent of the country's total energy requirement from this one source. Dunkerly et al., *Energy Strategies for Developing Nations*, give 53 trillion btu's as the crop and animal waste potential. A higher-income Senegal would consume more eggs, milk, poultry, and meat. Higher-yielding crops also would characterize a high-income society. More animals and more crop waste would bring the energy resource to 80–100 trillion btu's.

Urban wastes, using per capita figures from Western Europe but adjusted downward for more recycling of paper, come to 70–80 trillion btu's. Adding this to forestry figures and crop wastes gives the figures in the text.
28. Goldemberg et al., *An End-Use Oriented Global Energy Strategy*.

8   Some Notes on the Transition

1. This is the approach used by Amory Lovins. See Lovins et al., *Least-Cost Energy*. Dennis Hayes also has examined the question in this fashion and credits Theodore Taylor with suggesting the idea.
2. An energy-efficient United States (as defined in chapter 2) would use about 80 million btu's per capita per year. Japan and China, for reasons given in chapters 6 and 7, would use about 50 million. This is for fully industrialized economies at present standards of per capita consumption in the world's wealthiest nations. A world average of 60 million, with 8 billion people, leads to the estimate of 480 quads. Lovins (ibid.) carries efficiency gains so far in West Germany that per capita usage is 30 million btu's. That implies a world total of 240 quads, a little less than present usage.
3. International Institute for Applied Systems Analysis, *Energy in a Finite World*

(Cambridge, Mass.: Ballinger, 1981). Renewable sources net of conversion losses and without any allowance for photovoltaic electricity exceed 500 quads. Energy analysis is confused by a bewildering choice of units. That study uses as a unit the "terawatt," approximately 30 quads. Major sources, after conversion losses, are biomass (180 quads), hydroelectricity (90 quads), wind electricity (90 quads), solar heat (66 quads), and geothermal energy (60 quads).

4. Bent Sorensen, *Global Energy Policy and Development Strategy* (Copenhagen: Niels Bohr Institute, University of Copenhagen, 1979). An excellent and detailed study with a world perspective is that by Deudney and Flavin, *Renewable Energy—The Power to Choose*. They estimate long-run renewable energy potential at 318–367 quads (p. 256). Their estimates of direct solar energy, wind, and photovoltaic electricity are very conservative.

5. Its high-growth target for the industrial nations is 4.3 percent and for the developing nations, 5.5 percent. Weighting these rates by 1984 shares of world output gives the 4.6 percent average used in the text. World Bank, *World Development Report, 1985*. Growth in world output for the 1950–73 period has been estimated at 4.7 percent. Kravis et al., *World Product and Income*.

6. Implicit rates for the United States are about 2.8 percent for the next decade and 2.5 percent after that. The labor force will be growing more slowly than it has in recent years. Europe and Japan are even closer to stable populations. Overall growth rates are essentially per capita growth rates. Growth rates of 2.5 percent to 3 percent, given stable populations, double per capita income every 23–28 years.

7. If population growth rates can be brought down to 1 percent (on their way to zero), then output growth rates of 5–6 percent in the developing nations imply doubling per capita incomes every 12–17 years.

8. This approach was suggested by Robert Stobaugh and Daniel Yergin in a guest editorial for the *New York Times* (October 30, 1980). Their title was "Put Energy into a 3% Solution." After they noted the coming decline in U.S. oil production and the near-impossibility of meeting estimates for increased supply in the usual scenarios, they proposed that the United States aim for annual gains of 3 percent in energy efficiency. The United States has maintained a higher rate of gain since 1979, but that is a rather short period. Denmark and Japan have maintained similar or higher rates for a decade.

   With economic growth at a rate lower than the rate of energy-efficiency gain, total energy use would decline first in the United States and then in the rest of the industrial world. In the developing nations, growth rates of 5–7 percent with energy efficiency gains of 3–3.5 percent would imply increasing energy use. Some of this would be renewable energy and some would be oil and gas. The latter would be available, in part, from lessened demands in the industrial nations.

9. World energy use is running at a rate of about 280 quads, including the Soviet bloc nations. Department of Energy, Energy Information Administration, *Annual Energy Review, 1984* (Washington, D.C.: U.S. Government Printing Office, 1985). This means that new renewable energy sources of 1.4 to 2.8 quads a year would be needed.

10. IEA, *World Energy Outlook* (1982), p. 232.

11. The IEA's 1982 *World Energy Outlook* developed a hypothetical oil use profile using its (rather high) energy demand forecasts and its estimates of remaining world oil re-

sources. That profile showed oil output rising to 67 million barrels a day in 1990 (it is now about 53 million) and staying there *for thirty years*. It declines thereafter at about 1 percent per year for twenty-five years.

The agency's profile for natural gas shows increasing production until the year 2010, followed by a rapid or gradual decline, depending on one's estimate of ultimately recoverable reserves. Ibid., p. 215 and p. 365.

12. Department of Energy, *Annual Energy Review, 1984.* The study assumes that the rate of efficiency gain will decline to about 1 percent per year. With economic growth in the 3 percent range or more, it follows that energy use will increase and that oil imports will rise.

13. Japanese data are for 1973–83 and are taken from World Bank, *World Development Report,* various years. Danish data are from European Investment Bank, *Information,* May 1985.

14. Joji Watanuki, "Japanese Society and the Limits of Growth," in Yergin and Hillenbrand, eds., *Global Insecurity.*

15. *Economist,* July 6, 1985. World Bank figures are consistent with this conclusion (*World Development Report, 1985*).

16. World Bank, *The Energy Transition in Developing Countries.*

17. World hydroelectric output rose at a rate of .63 quads per year in the 1973–82 period. Worldwide development of direct solar energy at rates already attained in the United States, Japan, Australia, and Israel, and worldwide development of wind, geothermal, and photovoltaic sources at rates already attained in the United States, would add another .7 quads per year. The remainder is well within the capability of biomass resources, properly managed.

18. For the United States, prices have jumped and then declined twice. The mid-1985 price for imported crude oil was still about six times its 1973 price after adjusting for inflation. Europe has had yet a third cycle, since oil is priced in dollars and the dollar appreciated sharply in 1984.

Any commodity with low short-term supply and demand elasticities and higher long-run elasticities is likely to exhibit cyclical price behavior. That oil price cycles have not appeared in the past says a great deal about the degree of competition in past oil markets.

19. One such perspective is given by the economic theory of the price and quantity path through time for exhaustible resources. Abstracting from discovery and extraction costs, prices will rise through time at a rate equal to the rate of interest. The 1978–81 price jump went clearly beyond any long-run sustainable price trend.

At a less theoretical level, even modest supply elasticities for non-OPEC oil and modest demand elasticities would result in a falling market for OPEC oil. The effect would be magnified by the decline in world output resulting from the "oil price shock."

20. One such was S. Fred Singer in a *New Republic* article in January 1979. Others in 1974 had predicted a speedy disintegration of OPEC. That turned out to be a premature judgment.

21. *Monthly Energy Review,* March 1985.

22. North Sea estimates are by Philip K. Verleger, Jr., a Washington oil analyst. He is

quoted in *Business Week*, July 8, 1985. Alaskan oil must bear the marginal cost of pipeline and ship transport as well as high operating costs at the wellhead.

23. The estimate is made as follows: in a competitive oil market, capacity would be brought on line in ascending order of short-run marginal cost. OPEC has about 14–15 million barrels per day of unused capacity, of which 8–9 million barrels is in Saudi Arabia. All of that capacity presumably would be in use in competitive market conditions. USSR output presumably would be unaffected (nearly 12 million barrels), although exports might be. OPEC, the USSR, Mexico, and China could produce 48 million barrels so that the United States, Canada, the United Kingdom, and all other producers could operate only the lowest-cost facilities. Under present market conditions it would be a safe guess that $8–$10 per barrel represents an upper limit. These estimates are consistent with those given in the *New York Times*, February 9, 1986.

24. One of the most optimistic outlooks for the future relies heavily on these sources. See William M. Brown, "The Outlook for Future Petroleum Supplies" in Simon and Kahn, eds., *The Resourceful Earth*.

25. Choe, *A Model of World Energy Markets and OPEC Pricing*.

26. S. Fred Singer, *Wall Street Journal*, January 18, 1985, and July 5, 1985; *Washington Post*, March 9, 1986. Adelman is quoted in the *New York Times*, February 19, 1986.

27. The supporting calculation is in chap. 5, n. 8.

9   Getting There from Here: Some Policy Proposals

1. The Department of the Treasury's draft tax reforms released in 1984 would have eliminated all tax incentives for all forms of energy supply: depletion allowances in excess of cost, immediate charges to expense of all drilling costs, investment credits, energy investment credits, and accelerated depreciation. The tax-free reinvestment of utility dividends would expire.

    The president's version of tax reform reflected successful lobbying by the oil and gas industry; most of the tax preferences that Treasury proposed to eliminate reappeared. Some have been reduced as Congress considers tax reform. At this writing (1986) it appears that a tax reform bill will pass, that some tax benefits for conventional energy supply will remain, and that some renewable energy tax benefits will be restored.

2. National Research Council, Carbon Dioxide Assessment Committee, *Changing Climate* (Washington, D.C.: National Academy of Science, 1983). The Environmental Protection Agency released a report a few days before the NRC study. Both pointed to the uncertainty in models used to predict climate changes, but both regarded whatever changes might follow CO_2 buildup as inevitable. As an example of the impact of a rising sea level, the EPA report considered Charleston, S.C., and the billion dollar cost of a four- to seven-foot rise in sea level in that one community alone.

3. Lovins et al., *Least-Cost Energy*.

4. See, for example, *Energy: The Next Twenty Years*. The chapter on nuclear power covers these issues in a balanced fashion.

5. Methanol-fueled vehicles are much less polluting than present vehicles with respect to nitrogen compounds. Additional design work will be needed to ensure that organics are minimized; some tests to date show mixed results. EPA tests are cited in Lester R.

Brown et al., *State of the World, 1985* (New York: Norton, 1985), p. 195. This is a Worldwatch Institute volume that deals with a range of issues on population, resources, and the environment.

6. Markets can, in principle, deal with "split incentives." Energy-efficient homes, having lower fuel and electricity bills, should command higher prices. Builders or short-term owners should thus have financial incentives to buy energy-efficient features. Energy-efficient apartments or office buildings should likewise command higher rents (where tenants pay the bills) or bring higher net yields (where landlords pay the bills). These signals do not seem to work very well. Markets could, in principle, replace fire codes. Fire-prone buildings should command lower rents and bring lower returns. Still, we use fire codes—perhaps as a short-cut in obtaining information.

7. National Audubon Society, *The Audubon Energy Plan, 1984*, vol. 2, p. D-7.

8. Stobaugh and Yergin, eds., *Energy Future*, p. 202.

9. This program is proposed in SERI, *A New Prosperity*. The program would be less costly than the synfuels program and would bring major added benefits in helping to restore competitiveness to large sectors of basic industry in the United States.

10. This approach was proposed by the staff of the California Public Utilities Commission in establishing how much of the cost of the Palo Verde nuclear plant should be allowed into the rate base. Southern California Edison protested vigorously. The proposal has much merit for any proposed new facilities but can hardly be applied retroactively.

11. The most efficient refrigerators now on the market abroad, as chapter 2 pointed out, are about four times as efficient as U.S. 1980 models. A prototype reported to be still 2–4 times more efficient is cited in D. B. Goldstein, "Refrigerator Reform: Guidelines for Energy Gluttons," *Technology Review*, February–March 1983.

12. Unocal's problems in getting its Parachute, Colorado, plant to operate are described in *Investor's Daily*, August 5, 1985. The plant has been completed for two years, but one step in the process is not working. The article was based on an interview with the vice-chairman of the U.S. Synthetic Fuels Corporation.

13. U.S. 1984 oil consumption was 31 quads, of which 21 was domestically produced and 10 imported. Estimates of domestic output in the mid-1990s vary, but all show a decline. I follow *Energy Future* (Stobaugh and Yergin, p. 55) and use 15–16 quads as an estimate for 1990 production.

    If imports were to be held at the 1984 level with the economy growing at 2.8 percent a year, oil output declining, and energy needs to be met without recourse to more coal or nuclear power, it would appear that a miracle would be needed. Yet that could be accomplished with nothing more spectacular than things that already have happened. Continued efficiency gains at 3.5 percent per year, and renewable energy growth at 5 percent per year, would suffice to meet energy demands.

14. The higher-than-apparent marginal cost of imported oil is described and roughly calculated in ibid., p. 66.

15. Appliance standards are now in place in California (where they are being tightened) and in Maine. Active efforts are under way in Florida, New York, Oregon, and several other states.

16. Energy expenditures in 1985 were about $450 billion in a GNP of $4,000 billion. (Energy expenditures are estimated by taking Department of Energy Energy Information Administration data for 1984 and applying 1985 estimated prices and quantities.)

A low estimate of the required subsidy can be made by assuming perfectly elastic supply and applying a demand elasticity coefficient to see what price reduction would bring the quantity of energy demanded to 100 quads. This is a crude approach, since it uses a composite elasticity and a composite of energy forms. Better estimates would come from elaborate models with an input-output table and proper demand and supply elasticities for each fuel and each sector.

The crudely estimated figure, which is an underestimate, is so large as to make my point. No one would now seriously propose to do this. Earlier energy supply plans that called for large federal expenditures to make sure that "adequate" supplies of energy were available were, in fact, proposals to do the above—proposals that would not now be considered seriously.

17. Existing subsidies for conventional energy supply have been estimated in various ways. This figure includes tax deductions and credits, budget appropriations, grants, loans, and loan guarantees. It does not include any estimate of the value of implicit federal insurance of nuclear plants. Calculations by Richard Heede, Rocky Mountain Institute, reported in *Renewable Energy News*, August 1985, or *Wall Street Journal*, September 17, 1985.

Bibliography

Adelman, Morris A., et al. *No Time to Confuse* (San Francisco: Institute for Contemporary Studies, 1975).

Barnett, H. J., and C. Morse. *Scarcity and Growth: The Economics of Natural Resource Availability* (Baltimore: Johns Hopkins University Press for Resources for the Future, 1963).

Beckman, Petr. "Solar Energy and Other 'Alternative' Energy Sources," in Julian L. Simon and Herman Kahn, eds., *The Resourceful Earth* (Oxford: Basil Blackwell, 1984).

Brown, Lester R., et al. *State of the World, 1985* (New York: W. W. Norton, 1985).

Bupp, Irvin C., and Jean-Claude Derian. *The Failed Promise of Nuclear Power* (New York: Basic Books, 1981).

————. *Light Water: How the Nuclear Dream Dissolved* (New York: Basic Books, 1978).

California Energy Commission. *Securing California's Energy Future: 1983 Biennial Report* (Sacramento: California Energy Commission, 1983).

Capehart, Barney L., John F. Alexander, and Lynne C. Capehart. *Florida's Electric Future: Building Plentiful Supplies on Conservation* (Gainesville: University of Florida Center for Wetlands, 1982).

Charlier, Roger H. *Tidal Energy* (New York: Van Nostrand, 1982).

Choe, Boum-Jong. *A Model of World Energy Markets and OPEC Pricing*. World Bank Staff Working Papers, Number 633 (Washington, D.C.: World Bank, 1984).

Commission of the European Communities. *A European Energy Strategy* (Brussells: EEC, 1984).

Confederation Française Democratique du Travail, Groupe Confederal Energie. *Le Dossier de l'Energie* (Paris: Editions du Seuil, 1984).

Darmstadter, Joel, et al. *How Industrial Societies Use Energy* (Baltimore: Johns Hopkins University Press, 1977).

Department of Energy. *Annual Energy Review, 1984* (Washington, D.C.: U.S. Government Printing Office, 1985).

————. *Inventory of Power Plants, 1984* (Washington, D.C.: U.S. Government Printing Office, 1984).

Deudney, Daniel, and Christopher Flavin. *Renewable Energy—The Power to Choose* (New York: W. W. Norton, 1983).

Dunkerley, Joy, et al. *Energy Strategies for Developing Nations* (Baltimore: Johns Hopkins University Press for Resources for the Future, 1981).

Energy: The Next Twenty Years. A Report of the Study Group Sponsored by the Ford Foundation and Administered by Resources for the Future (Cambridge, Mass.: Ballinger, 1979).

Energy Policy Project of the Ford Foundation. *A Time to Choose* (Cambridge, Mass.: Ballinger, 1974).

Energy Research and Development Administration. *A National Plan for Energy Research, Development and Demonstration: ERDA 48* (Washington, D.C.: U.S. Government Printing Office, 1975).

Flavin, Christopher. *Renewable Energy at the Crossroads* (Washington, D.C.: Center for Renewable Resources, 1985).

————. *Wind Energy: A Turning Point* (Washington, D.C.: Worldwatch Institute, 1981).

Gabel, Medard. *Energy, Earth and Everyone* (New York: Audubon Books, 1980).

Gas Research Institute, *Methane from Biomass and Wastes Research* (Chicago: Gas Research Institute, 1984).

Gibbons, John P., and William U. Chandler. *Energy: The Conservation Revolution* (New York: Plenum, 1981).

Goldemberg, Jose, et al. *An End-Use Oriented Global Energy Strategy*, Report #179 (Princeton, N.J.: Princeton Center for Energy and Environmental Studies, 1985).

Goodwin, Craufurd D., ed. *Energy Policy in Perspective* (Washington, D.C.: The Brookings Institution, 1981).

Hayes, Denis. *Rays of Hope: The Transition to a Post-Petroleum World* (New York: W. W. Norton, 1977).

Hollander, Jack M., ed. *Annual Review of Energy, 1983* (Palo Alto, Calif.: Annual Reviews, Inc, 1983).

International Energy Agency. *World Energy Outlook* (Paris: OECD, 1977).

————. *World Energy Outlook, 1982* (Paris: OECD, 1982).

International Institute for Applied Systems Analysis (IIASA). *Energy in a Finite World* (Cambridge, Mass.: Ballinger, 1981).

Katz, Rosamund, et al. *Energy in New England: Transition to the 1980's* (Washington, D.C.: New England Congressional Institute, 1981).

Kendall, Henry W., and Steven J. Nadis, eds. *Energy Strategies: Toward a Solar Future* (Cambridge, Mass.: Ballinger, 1980).

Kohl, Wilfred L., ed. *After the Second Oil Crisis: Energy Policies in Europe, America and Japan* (Lexington, Mass.: D. C. Heath, 1982).

Kravis, Irving B., et al. *World Product and Income* (Baltimore: Johns Hopkins University Press, 1982).

Lipschutz, Ronnie D. *Radioactive Waste: Politics, Technology and Risk* (Cambridge, Mass.: Ballinger, 1980).

Lovins, Amory. *Soft Energy Paths* (San Francisco: Friends of the Earth, 1977).

————, and L. Hunter Lovins. *Brittle Power: Energy Strategy for National Security* (Andover, Mass.: Brick House, 1982).

————. et al. *Least-Cost Energy: Solving the CO₂ Problem* (Andover, Mass.: Brick House, 1981).

Luneau, M., *Les Energies Nouvelles: Qu'en Espérer?* (Paris: La Documentation Française, 1982).

Maycock, Paul D., and Edward N. Stirewalt. *Photovoltaics* (Andover, Mass.: Brick House, 1981).

————. *A Guide to the Photovoltaic Revolution* (Emmaus, Pa.: Rodale Press, 1985).

National Audubon Society. *The Audubon Energy Plan, 1984* (New York: National Audubon Society, 1984).

National Conference on Renewable Energy Technologies: Proceedings (Honolulu: University of Hawaii, 1980).

National Research Council. *Energy in Transition, 1985–2010.* Final Report of the Committee on Nuclear and Alternative Energy Systems (San Francisco: W. H. Freeman, 1979).

————, Carbon Dioxide Assessment Committee. *Changing Climate* (Washington, D.C.: National Academy of Sciences, 1983).

————, Committee on Nuclear and Alternative Energy Systems. *U.S. Energy Supply Prospects to 2010* (Washington, D.C.: National Academy of Sciences, 1979).

New York City Energy Office. *Energy Consumption in New York City* (New York: New York City Energy Office, 1981).

Northwest Power Planning Council. *Regional Conservation and Electric Power Plan, 1983* (Portland, Oreg.: Northwest Power Planning Council, 1983).

Office of Technology Assessment. *Energy from Biological Processes* (Washington, D.C.: U.S. Government Printing Office, 1980).

————. *Industrial Energy Use* (Washington, D.C.: U.S. Government Printing Office, 1983).

————. *U.S. Vulnerability to an Oil Import Curtailment* (Washington, D.C.: U.S. Government Printing Office, 1984).

Olivier, David, and Hugh Miall. *Energy-Efficient Futures: Opening the Solar Option* (London: Earth Resources Research, Ltd., 1983).

Oregon Alternate Energy Development Commission. *Future Renewable.* Final Report (Salem: State of Oregon, 1980).

Organization for Economic Co-operation and Development. *Energy Balances of OECD Countries, 1973–1975* (Paris: OECD, 1976).

————. *Energy Balances of OECD Countries, 1982–1983* (Paris: OECD, 1985).

Piermont, Laurent. *L'énergie Verte* (Paris: Editions du Seuil, 1982).

Report of the Demand and Conservation Panel to the Committee on Nuclear and Alternative Energy Systems (CONAES). *Alternative Energy Demand Futures to 2010* (Washington, D.C.: National Research Council, National Academy of Sciences, 1979).

Ross, Marc H., and Robert H. Williams. *Our Energy: Regaining Control* (New York: McGraw-Hill, 1981).

Sant, Roger W. *The Least-Cost Energy Strategy* (Arlington, Va.: The Energy Productivity Center, Mellon Institute, 1979).

————, et al. *The Least-Cost Energy Strategy, 1978–2000* (Arlington, Va.: The Energy Productivity Center, Mellon Institute, 1981).

Sawhill, John C., ed. *Energy Conservation and Public Policy* (Englewood Cliffs, N.J.: Prentice-Hall, 1979).

Solar Energy Research Institute (SERI). *A New Prosperity: Building a Sustainable Energy Future* (Andover, Mass.: Brick House, 1981).

Sorensen, Bent. *Global Energy Policy and Development Strategy* (Copenhagen: Niels Bohr Institute, University of Copenhagen, 1979).

Steinhart, John S., et al. *Pathway to Energy Sufficiency* (San Francisco: Friends of the Earth, 1979).

Stobaugh, Robert, and Daniel Yergin, eds. *Energy Future.* Report of the Energy Policy Project at the Harvard Business School (New York: Random House, 1979).

Taylor, Vince. *Energy: The Easy Path* (Cambridge, Mass.: Union of Concerned Scientists, 1979).

United Nations. *Economic and Social Survey of Asia and the Pacific, 1980* (New York: United Nations, 1981).

El Vent: Energia Eòlica a Catalunya (Barcelona: Gèneralitat de Catalunya, 1984).

Workshop on Alternative Energy Strategies. *Energy: Global Prospects 1985–2000* (New York: McGraw-Hill, 1977).

World Bank. *The Energy Transition in the Developing Countries* (Washington, D.C.: World Bank, 1983).

————. *Renewable Energy Resources in the Developing Countries* (Washington, D.C.: World Bank, 1980).

————. *World Development Report, 1985* (Washington, D.C.: World Bank, 1985).

World Energy Conference. *Survey of Energy Resources* (New York: U.S. National Committee of the World Energy Conference, 1974).

Yergin, Daniel, and Martin Hillenbrand, eds. *Global Insecurity* (Boston: Houghton Miflin, 1982).

Index

Library of Congress Cataloging-in-Publication Data
Blackburn, John O., 1929–
The renewable energy alternative.
Bibliography: p.
Includes index.
1. Renewable energy sources. I. Title.
TJ808.B57 1987 333.79 86-29273
ISBN 0-8223-0687-5
ISBN 0-8223-0744-8 (pbk.)